ACS SYMPOSIUM SERIES **863**

Pesticide Decontamination and Detoxification

Jay J. Gan, EDITOR
University of California at Riverside

Peter C. Zhu, EDITOR
Johnson and Johnson

Steven D. Aust, EDITOR
Utah State University

Ann T. Lemley, EDITOR
Cornell University

Sponsored by the
ACS Division of Environmental Chemistry, Inc.

American Chemical Society, Washington, DC

Library of Congress Cataloging-in-Publication Data

Pesticide decontamination and detoxification / Jay J. Gan, editor ...[et al.] ; sponsored by the ACS Division of Environmental Chemistry, Inc.

  p. cm.—(ACS symposium series ; 863)

  Includes bibliographical references and index.

  ISBN 0-8412-3847-2

  1. Pesticides—Environmental aspects—Congresses. 2. Pesticides—Biodegradation—Congresses

  I. Gan, Jay J., 1963- II. American Chemical Society. Division of Environmental Chemistry, Inc. III. American Chemical Society. Meeting (224th : 2002 : Boston, Mass.) IV.Series.

TD196.P38P462   2003
628.5′29—dc21   2003056091

The paper used in this publication meets the minimum requirements of American National Standard for Information Sciences—Permanence of Paper for Printed Library Materials, ANSI Z39.48-1984.

Copyright © 2004 American Chemical Society

Distributed by Oxford University Press

All Rights Reserved. Reprographic copying beyond that permitted by Sections 107 or 108 of the U.S. Copyright Act is allowed for internal use only, provided that a per-chapter fee of $24.75 plus $0.75 per page is paid to the Copyright Clearance Center, Inc., 222 Rosewood Drive, Danvers, MA 01923, USA. Republication or reproduction for sale of pages in this book is permitted only under license from ACS. Direct these and other permission requests to ACS Copyright Office, Publications Division, 1155 16th St., N.W., Washington, DC 20036.

The citation of trade names and/or names of manufacturers in this publication is not to be construed as an endorsement or as approval by ACS of the commercial products or services referenced herein; nor should the mere reference herein to any drawing, specification, chemical process, or other data be regarded as a license or as a conveyance of any right or permission to the holder, reader, or any other person or corporation, to manufacture, reproduce, use, or sell any patented invention or copyrighted work that may in any way be related thereto. Registered names, trademarks, etc., used in this publication, even without specific indication thereof, are not to be considered unprotected by law.

PRINTED IN THE UNITED STATES OF AMERICA

Foreword

The ACS Symposium Series was first published in 1974 to provide a mechanism for publishing symposia quickly in book form. The purpose of the series is to publish timely, comprehensive books developed from ACS sponsored symposia based on current scientific research. Occasionally, books are developed from symposia sponsored by other organizations when the topic is of keen interest to the chemistry audience.

Before agreeing to publish a book, the proposed table of contents is reviewed for appropriate and comprehensive coverage and for interest to the audience. Some papers may be excluded to better focus the book; others may be added to provide comprehensiveness. When appropriate, overview or introductory chapters are added. Drafts of chapters are peer-reviewed prior to final acceptance or rejection, and manuscripts are prepared in camera-ready format.

As a rule, only original research papers and original review papers are included in the volumes. Verbatim reproductions of previously published papers are not accepted.

ACS Books Department

Contents

Preface..xi

Biologically Based Mechanisms

1. Detoxification and Metabolism of Chemicals
   by White-Rot Fungi..3
   Steven D. Aust, Paul R. Swaner, and James D. Stahl

2. Microbial and Photolytic Degradation of 3,5,6-
   Trichloro-2-pyridinol...15
   Yucheng Feng

3. Biological Detoxification of Organophosphate Pesticides...................25
   Mark Shimazu, Wilfred Chen, and Ashok Mulchandani

4. Evolution of New Enzymes and Pathways: Soil Microbes
   Adapt to s-Triazine Herbicides...37
   Lawrence P. Wackett

Chemically Based Mechanisms

5. Detoxification of Some Halogenated Pesticides
   by Thiosulfate Salts...51
   J. Gan and S. Bondarenko

6. Anodic Fenton Degradation of Pesticides...65
   A. T. Lemley, Q. Wang, and D. A. Saltmiras

7. Chemical Detoxifying Neutralization of *ortho*-
   Phthalaldehyde: Seeking the "Greenest"..85
   Peter C. Zhu, Charles G. Roberts, and Jiejun Wu

8. Electrochemical Destruction of Triclosan..........99
   James Farrell, Jiankang Wang, and Ronald LeBlanc

9. Detoxification of Pesticide in Water Using Solar Photocatalysis.......113
   S. Malato and A. Agüera

Field Processes and Applications

10. Microbial Degradation of Atrazine in Soils, Sediments, and Surface Water..........129
    Mark Radosevich and Olli H. Tuovinen

11. Bioremediation of Atrazine-Contaminated Soil..........141
    Edward Topp, Fabrice Martin-Laurent, Alain Hartmann, and Guy Soulas

12. Detoxification of Pesticide Residues in Soil Using Phytoremediation..........155
    J. B. Belden, B. W. Clark, T. A. Phillips, K. L. Henderson, E. L. Arthur, and J. R. Coats

13. Remediation of Halogenated Fumigant Compounds in the Root Zone by Subsurface Application of Ammonium Thiosulfate..........169
    Sharon K. Papiernik, Frederick F. Ernst, Robert S. Dungan, Wei Zheng, Mingxin Guo, and Scott R. Yates

14. Detoxification and Destruction of PCBs, CAHs, CFCs, and Halogenated Biocides in Soils, Sludges, and Other Matrices Using Na/NH_3..........181
    Charles U. Pittman, Jr.

15. Comparison of Atrazine and Alachlor Sorption, Mineralization, and Degradation Potential in Surface and Aquifer Sediments..........199
    Sharon A. Clay, David E. Clay, and Thomas B. Moorman

16. Pesticide Runoff and Mitigation at a Commercial Nursery Site..........213
    J. N. Kabashima, S. J. Lee, D. L. Haver, K. S. Goh, L. S. Wu, and J. Gan

17. Impacts of Surfactant Adjuvants on Pesticide Availability
    and Transport in Soils..231
    Kurt D. Pennell, Ahmet Karagunduz, and Michael H. Young

Indexes

Author Index..348

Subject Index...349

Preface

The 224th American Chemical Society (ACS) National Meeting and Exposition in Boston, Massachusetts included a program organized by Peter C. Zhu of Johnson & Johnson entitled *Deactivation (Neutralization or Detoxification) and Safe Disposal of Germicides and Pesticides*. The program consisted of 18 presentations plus this Introduction and Concluding Remarks by Ann T. Lemley of Cornell University. The program was remarkable because it included presentations covering a wide variety of chemicals, most matrices, and many treatment systems, several of which were very new and innovative (*see* Table 1). Examples of treatment systems addressed include everything from some rather standard microbial and chemical treatment systems to new and innovative systems such as thiosulfate dehalogenation (Chapter 13 by S. K. Papiernick et al.), electrochemical destruction (Chapter 8 by J. Farrell et al.), including anodic Fenton-based degradation, (Chapter 6 by A. T. Lemley et al.), and phytoremediation (Chapter 12 by J. B. Blenden et al.).

The list of chemicals addressed in the presentations included general classes of chemicals, such as organophosphates, chlorinated aromatic hydrocarbons, petroleum hydrocarbons and other pesticides, herbicides, biocides, fumigants, and germicides, but some presentations addressed specific chemicals such as Atrazine, Amitraz, and Bifenthrin. Several presentations described the degradation of many chemicals that were previously considered quite persistent or resistant to degradation. Chemical degradation in almost every matrix, except perhaps air, was addressed. These matrices included both surface and subsurface soils, surface, runoff and wash water, sludges, sediments, and so on.

The program seemed remarkable when considering the evolution of this subject area. Not many years ago, ACS meetings included presentations that simply described very serious cases of environmental pollution. It seemed clear that our water, air, and soils were incredibly contaminated. Considerable effort was devoted to analytical methods, frequently to document the seriousness of the environmental pollution. It

Table 1. Chemicals, Treatment Systems and Matrices addressed in the ACS Symposium

Chemicals	Chlorinated pyridinal Organophosphate pesticides *o*-Phthalaldehyde Triclosan Polychlorinated biphenyls Chlorofluorocarbons Chlorinated aromatics hydrocarbons Halogenated biocides Atrazine Petroleum hydrocarbons Amitraz Bifenthrin Other pesticides, herbicides, biocides, fumigants, germicides, etc.
Treatment Systems	White-rot fungi Microbial Photolytic Enzymes Thiosulfates Anodic Fenton Electrochemical Ultrasonic Chemical reduction Chemical oxidation Phytoremediation Solar photocatalysis
Matrices	Soil (surface and subsurface) Water (surface, runoff, vats, etc.) Sludges (solutions or sludges) Sediments Other

should be pointed out that the analytical chemists were successful in developing analytical methods for many chemicals in all kinds of matrices. However, they painted a rather gloomy picture of environmental pollution. More and more environmental pollution sites were being found, sometimes resulting in criticism of the ability of analytical chemists to come up with analytical methods that are too sensitive. These methods detected environmental pollutants in more and more sites, and encouraged the setting of lower and lower standards by the Environmental Protection Agency.

Now, it would seem that the next group of scientists, including chemists and biologists and others interested in remediation, are equally up to their task, or "the challenge". It would seem that many scientists only needed to be informed of the problem because their abilities are equal to the challenge. A myriad of systems has been developed to treat more and more recalcitrant chemicals in more and more matrices. Although it may not be time to rest, it's obviously time to recognize the accomplishments of the scientists who have responded actually quite quickly to a serious need, which has resulted in a noticeably better environment. The "war" is not over, but the arsenal has been vastly improved by very innovative and capable scientists. At the same time, a shift in the chemical industry to produce and use less persistent and more selective chemicals has occurred.

Not all of the ACS program have been included in this book, which was edited by Jay J. Gan, Peter C. Zhu, Ann T. Lemley, and myself. Contributions from additional authors, who could not participate in the symposium but whose work contribute to an even more inclusive treatise, are included in this book. One particular contribution, Chapter 4 by Lawrence P. Wackett entitled "Evolution of New Enzymes and Pathways: Soil Microbes Adapt to *s*-Triazine Herbicides" is especially appropriate because it concerns our ability to respond to the environmental pollution problem. This chapter describes a *Pseudomonas* species as a model to understand how bacterial genes involved in the metabolism of anthropogenic chemicals may arise and spread in the environment. The author describes how the ability to metabolize Atrazine by a *Pseudomonas* species evolved by recruiting enzymes from the amidohydrolase superfamily to form a metabolic pathway to efficiently metabolize *s*-triazine herbicides. Examples of the bioremediation of Atrazine-contaminated soil are described by Edward Topp, et al. in Chapter 11.

Detoxification by an additional method not covered in the symposium, the solar photocatalytic detoxification of pesticides in water using TiO_2, is described by Malato and Agüera in Chapter 9.

In summary, this book, resulting from an ACS program organized by Peter C. Zhu (and additional contributions resulting from efforts by Jay Gan) documents the fact that exceptional progress can result when the efforts of good scientists are applied to a problem. Sometimes the problem toward which their effort needs to be focused has only to be pointed out, but their accomplishments should not go unrecognized. When the abilities of chemists and biologists are focused on a particular problem the results can be impressive. The efforts of the biologists are synergized by the adaptability of microorganisms. Given the abilities of molecular biologists along with the recognition of how enzymes evolved by consolidation or incorporation of genes or gene fragments for appropriate motifs, one might expect even more progress through molecular engineering. Genes can now be synthesized and superior enzymes produced completely in vitro. Engineered biological systems can include numerous advantages. Most environmental pollution sites should not be left for "natural attenuation". It makes one wonder how exciting future ACS programs on this subject might be. We are in a biological, chemical, and technological revolution that hopefully will be applied to, or in recognition of, our environment.

Steven D. Aust
Department of Chemistry and Biochemistry
Utah State University
Logan, UT 84322-4705
435-797-2730 (telephone)
435-797-2755 (fax)
sdaust@cc.usu.edu

Biologically Based Mechanisms

Chapter 1

Detoxification and Metabolism of Chemicals by White-Rot Fungi

Steven D. Aust[*], Paul R. Swaner, and James D. Stahl

Chemistry and Biochemistry Department, Biotechnology Center,
Utah State University, Logan, Utah 84322–4705
[*]Corresponding author: email: sdaust@cc.usu.edu

White-rot fungi can degrade a wide variety of environmental pollutants using a variety of extracellular enzymes and chemicals normally involved in lignin degradation. Examples of toxic chemicals shown to be degraded by white-rot fungi include pentachlorophenol, trinitrotoluene, trichloroethylene, cyanide and polyaromatic hydrocarbons. Pentachlorophenol is methylated by a transmembrane methyl transferase. Trinitrotoluene is reduced by a transmembrane redox potential associated with a proton pump that the fungus uses to establish a rather low (4.5) extracellular pH. Trichloroethylene is aerobically dechlorinated by peroxidases using the carboxylate anion radical. The peroxidases oxidize either veratryl alcohol or manganese which oxidize oxalic acid to form the carboxylate anion radical for reductive dechlorinations. Other chemicals can either be directly or indirectly oxidized to radicals by the peroxidases. In all cases the chemicals are detoxified such that relatively high concentrations of these chemicals can be degraded. In addition, sites contaminated with multiple chemicals, including these toxic chemicals, can be remediated.

© 2004 American Chemical Society

Environmental pollution is a global problem. Contaminating chemicals are or can become toxic, with potential of detrimental effects to the health and safety of people and the environment. Contaminated sites often contain a mixture of environmentally persistent compounds. The persistent nature of many pollutants contributes to the potential risks and difficulty of remediation. Bioremediation is a promising and potentially cost effective strategy to remediate these sites. A potential problem is that many of the compounds are toxic to microorganisms that otherwise might be able to degrade these chemicals. Therefore, toxicity frequently limits natural attenuation by microorganisms, especially in areas of high concentrations. Examples of toxic chemicals that appear in various pollution sites are explosives such as trinitrotoluene, halocarbons such as trichloroethylene, cyanide and polycyclic aromatic hydrocarbons, all of which have been shown to be degraded by white-rot fungi. In many cases toxic chemicals were actually used to inhibit microorganisms. A classic example is the extensive use of pentachlorophenol and creosote as wood preservatives to inhibit wood-rotting fungi. Sites contaminated with these chemicals must now be remediated and this might be accomplished by the very organisms which were intended to be controlled by these chemicals, as these compounds have also been shown to be degraded by the fungi.

Our laboratory showed in 1985 that the white-rot fungus *Phanerochaete chrysosporium* could mineralize a wide variety of structurally unrelated environmental pollutants, some of which are very toxic (1). It was demonstrated that most significant degradation was under ligninolytic conditions, induced by nutrient limitation. These fungi are most effective in degrading lignin when other simple nutrients are not available. As a consequence the lignin-degrading system of *P. chrysosporium* has been studied extensively and the fungus has also been used in many studies of chemical bioremediation. The lignin degradation systems of other wood-rotting fungi, including *Pleurotus eryngii* (2), *Trametes versicolor* (3), *Pleurotus ostreatus* (4), *Phelbia radiata* (5), *Ceriporiopsis subvermispora* (6), and *Bjerkandera* species (7), have been examined in many studies. These and other wood-rotting fungi are being extensively investigated for their use in what is now termed mycoremediation (8, 9).

Lignin

White-rot fungi are the only organisms known with the ability to depolymerize and mineralize lignin. This system has evolved to accommodate the unique chemical nature of lignin. Unlike most biopolymers, lignin is non-repeating and irregular. The biosynthesis of lignin is the result of peroxidase-catalyzed polymerization of three precursor alcohols: coumaryl alcohol, coniferyl alcohol, and synapyl alcohol (10, 11). These phenylpropanoid compounds are oxidized by peroxidases to generate radicals that combine randomly, forming interunit carbon-carbon and ether bonds resulting in an irregular polymer (12, 13). The relative starting concentration of each monomer is dependent upon plant species and cell type, thus the final structure varies greatly between species. The unique composition of lignin, being large, nonlinear, three-dimensional, non-hydrolyzable, and stereo-irregular, makes its degradation a complicated process that by necessity is extracellular and nonspecific.

Reaction Possibilities

The extracellular, free-radical, nonspecific, and partially redundant mechanisms of the lignin-degrading systems of white-rot fungi facilitate degradation of a wide variety of environmental pollutants. Fundamental to this ability is the formation of both oxidative and reductive species. The extracellular enzymes of white-rot fungi involved in the production of free-radicals include lignin peroxidases (LiP), manganese peroxidases (MnP), cellobiose dehydrogenase (CDH), and laccases. Membrane associated methyl transferases and a plasma membrane redox potential are also thought to contribute to the detoxification and degradation of compounds. Other factors related to free-radical generation of the wood–degrading system include hydrogen peroxide (H_2O_2) and H_2O_2 producing enzymes, small molecule mediators, such as veratryl alcohol (VA) and Mn(II), and oxalic acids. The free-radical reactions generated by these systems lead to both direct and indirect (mediated) oxidations and reductions.

Peroxidases

Lignin peroxidases have been shown to directly oxidize lignin (14) and a variety of bulky substrates (15), but it is the indirect redox reactions catalyzed by these enzymes that are considered important for the degradation of lignin and environmentally persistent pollutants (9). The peroxidase catalytic cycle is shown in figure 1A. The ferric form of the enzyme is oxidized by H_2O_2 generating the enzyme intermediate termed compound I. The oxidized enzyme can then oxidize two sequential, one-electron direct oxidations of reductants that can interact with the peroxidase, forming compound II and then regenerating ferric enzyme in the cycle (16). Although lignin peroxidases are strong oxidants with a reduction potential of 1.4 V, which can oxidize a variety of substrates, veratryl alcohol (VA) is a physiological substrate. Vertrayl alcohol is synthesized and secreted by the fungus under secondary metabolism and is a byproduct of lignin degradation. When VA is oxidized, it is converted to the veratryl alcohol cation radical ($VA^{\bullet+}$), another powerful oxidant. The $VA^{\bullet+}$ has been shown to mediate the oxidation of other chemicals, but it is not thought to diffuse away from LiP due to its limited stability in water (17).

Unlike LiP, MnP does not oxidize lignin or large molecules but is dependent upon manganese to complete its catalytic cycle (Figure 1A). Manganese peroxidase oxidizes chelated Mn(II) to Mn(III), which is believed to act as a diffusible redox mediator. It is believed that oxalic acid is the physiological chelator (18). The relatively stabalized Mn(III)-oxalic acid complex is then able to diffuse away from the enzyme and oxidize less accessible areas of wood or other chemicals (19).

The primary oxidations catalyzed by peroxidases can lead to secondary reactions that produce reductive radicals. Oxalic acid is produced by fungi and found in wood and very susceptible to oxidation. Both $VA^{\bullet+}$ and Mn(III) can oxidize oxalic acid resulting in homolytic cleavage and the formation of the carboxylate anion radical ($CO_2^{\bullet-}$) and CO_2 (20). The $CO_2^{\bullet-}$ has a reduction potential of -1.9 V and can reduce O_2 and other chemicals (21).

Laccase

Laccases are copper containing oxidases that catalyze the one electron oxidation of phenolic substrates (Figure 1B) with the concurrent reduction of molecular oxygen to water. Substrates include relatively easily oxidized aromatic compounds such as polyphenols, aromatic amines, and methoxy-substituted monophenols (22). Additionally, indirect oxidations are thought to occur in the presence of a suitable redox mediator such as 3-hydroxyanthranilate (3-HAA) (23).

Cellobiose Dehydrogenase

Cellobiose dehydrogenase (CDH) preferentially oxidizes cellobiose, the product of cellulose hydrolysis by cellulases. A variety of substrates can be used as electron acceptors, including quinones, transition metals, phenoxyl radicals, and O_2 (Figure 1C, 14). A possible secondary reaction is the production of Fenton reagent resulting in the hydroxyl radical.

Some potential substrates for CDH such as quinones and phenoxyl radicals are oxidation products of peroxidases and laccases formed during lignin degradation. This suggests the possibility of futile redox cycling where products of oxidation are simply reduced by CDH to produce the original substrate. It has

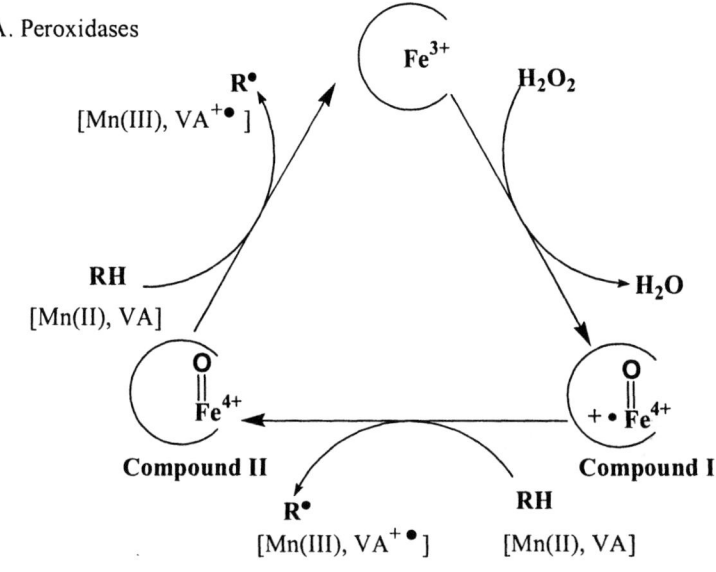

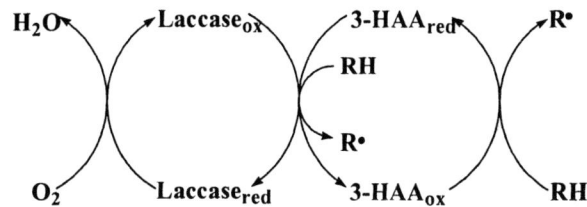

$H_2O_2 + Fe^{2+} \longrightarrow Fe^{3+} + OH^- + {}^\bullet OH$

${}^\bullet OH + Oxalate \longrightarrow OH^- + CO_2 + CO_2^{\bullet -}$

Figure 1. Direct and mediated oxidation and reduction reactions catalyzed by enzymes from white-rot fungi.

been suggested that reactions catalyzed by CDH facilitate degradation of lignin and cellulose instead of creating a futile redox cycle. Futile cycling of phenols and quinones might be circumvented by combing the action of CDH and a methyl transferase, as described below for the degradation of pentachlorophenol. A quinone or semiquinone or even a phenoxyl radical might be reduced by CDH, then methylated such that the aromatic ring can be oxidized by the peroxidases. Lignin degradation is also thought to be enhanced by CDH, by minimizing the repolymerization of phenoxyl radicals produced during the oxidation of lignin.

Pentachlorophenol Detoxification and Degradation

Pentachlorophenol (PCP) has been used extensively as a wood preservative, fungicide, bactericide, herbicide, etc. (24). Pentachlorophenol, like some other polychlorinated phenols, is a potent inhibitor of oxidative phosphorylation, thus highly toxic to microorganisms. Although PCP is generally thought to be toxic to microorganisms at concentrations over 50 ppm it was used at much higher concentrations. As a result there are many sites contaminated with very high concentrations of PCP. These high concentrations are not biodegraded naturally most likely due to their toxicity to microorganisms. However we have been able to show degradation of very high concentrations (i.e., 1600 ppm) of PCP by *P. chrysosporium* in soil (25).

White-rot fungi have a methylation system that is important in the degradation of phenolic pollutants and phenolic lignin degradation products (25, 26). Pentachloroanisole (PCA) was found upon analysis of extracts from cultures of *P. chrysosporium* degrading PCP (25). Since PCA is less toxic than PCP, this methylation is a very important detoxification step that was proposed to allow the fungus to degrade relatively high concentrations of PCP.

Methylation occurs extracellularly by a membrane-bound methyl transferase (Figure 2A). The physiological methyl donor is proposed to be S-adenosylmethionine, or methyl chloride, which is synthesized by white-rot fungi from S-adenosylmethionine (27). Methylation occurs under both ligninolytic and non-ligninolytic conditions. Thus, under non-ligninolytic conditions PCA accumulates. The ability to methylate and thus resist the toxicity of PCP is directly dependent upon the amount of fungal mycelia. Therefore it is important to provide sufficient fungus when mixing with PCP contaminated soil or water to provide resistance to the toxicity so that bioremediation can start. Methylation of PCP is also important for biodegradation because the aromatic ring is oxidized by the peroxidases instead of the phenol, resulting in dechlorination, further detoxification, and eventually ring opening and mineralization (Figure 2B).

It is believed that the resulting hydroquinones are methylated by a trans-membrane methyltransferase (28). This mechanism is also believed to limit the

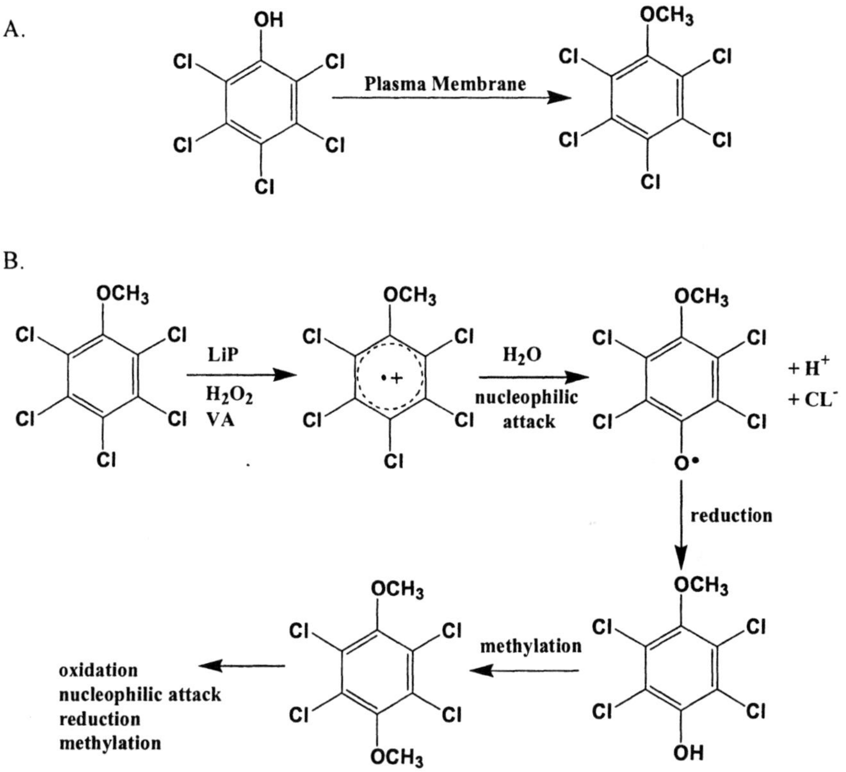

Figure 2. (A) Methylation of pentachlorophenol to pentachloroanisole by membrane bound methyltransferase from <u>Phanerochaete chrysosporium</u>. (B) Cycle of oxidative, reductive, and methylation reactions leading to the degradation of pentachloroanisole.

ability of quinones to redox cycle, providing a mechanism for degradation instead of redox cycling (29). Methylation of hydroquinones may also be involved in their mineralization.

Trinitrotoluene Detoxification and Degradation

Trinitrotoluenene (TNT) is another example of a toxic chemical that is involved in many sites that must be remediated. TNT is also very toxic to most microorganisms, at about 50 ppm (30). However, we showed that very high concentrations of TNT (i.e., 10,000 ppm, in laboratory microcosms) were also degraded by *P. chrysosporium* (31). Analysis of the culture medium during degradation of ^{14}C-TNT revealed that most if not all of the radioactivity remained outside the cells but it was rapidly converted to aminodinitrotoluenes, although some diaminonitrotoluene could be found after prolonged incubation (31). Since fungi were known to have a plasma-membrane associated redox potential (32) we proposed that this could be involved in the reduction of TNT to amine metabolites (33). This membrane redox potential was found to be responsible for detoxification of TNT (34). As was the case for the membrane methyltransferase, the redox potential is present in both ligninolytic and non-ligninolytic mycelia such that pregrowth of the fungus, to provide more of the plasma membrane redox potential, results in rapid reduction and detoxification such that very high concentrations of TNT can be degrade in soil or water (25, 31, 35).

We propose that the plasma membrane redox potential may be important in the metabolism of several chemicals (Figure 3, 36). We first showed that it could reduce and therefore detoxify ferricyanide (36), a toxin that is found in many pollution sites, such as town gas sites. The plasma membrane redox potential may also reduce toxic quinones, resulting from the oxidation of lignin and other aromatic chemicals by the peroxidases (36). In this case the resulting hydroquinones could also be methylated by the methyltransferases, similarly to PCP, to provide substrates for the peroxidases. These methoxyquinones would be oxidized by the peroxidases to products that will lead to their mineralization. The oxidation of hydroquinones by the peroxidase would result in redox cycling rather than ring opening and mineralization. Thus, the plasma membrane potential and methyltransferase, along with the peroxidases, could result in complete extracellular oxidation (mineralization) of these types of potentially toxic phenols and quinones. Physiologically, this redox potential is probably also important for the protection of fungal hyphae from free radicals produced by peroxidases or during lignin breakdown.

Trichloroethylene Dehalogenation and Degradation

Halogenated hydrocarbons such as trichloroethylene are also potentially toxic environmental pollutants (37). They are frequently metabolized by microbes to toxic and mutagenic metabolites. Dehalogentaion of halocarbons is possible although it usually requires anaerobic conditions (38). We proposed that halocarbons might be dehalogenated aerobically by white-rot fungi. We found that ligninolytic, aerobic cultures of *P. chyrsosporium* were able to mineralize trichloroethylene (TCE) and carbon tetrachloride (CCl_4) (39). The process turned out to be fairly complicated (Figure 4). The trichloromethyl radical was formed when lignin peroxidase was incubated with CCl_4, H_2O_2, veratryl alcohol and EDTA. We proposed that EDTA was reducing the veratryl alcohol cation radical to form $CO_2^{\bullet-}$, which is an excellent reductant with sufficient potential to reductively dechlorinate CCl_4. *P. chrysosporium* synthesizes and secretes oxalic acid, which is also in fairly high concentrations in rotting wood. We then showed that oxalic acid could replace EDTA in dechlorinating CCl_4 and suggested that oxalic acid was the physiological source of reducing equivalents. Oxalic acid is actually a di-alpha-keto acid which can be readily oxidatively decarboxylated to give CO_2 and $CO_2^{\bullet-}$, an excellent reductant. We subsequently showed that oxalic acid can also be oxidized by Mn(III), produced by the manganese-dependent peroxidases, to catalyze reductions (39). These are additional examples of veratryl alcohol and manganese serving as mediators for lignin and manganese-dependent peroxidases, respectively, to metabolize chemicals. In these cases they are mediating reductions, specifically reductive dechlorinations of halocarbons, using oxalic acid as a source of reducing equivalents.

Summary

A system has evolved to give white-rot fungi the ability to degrade lignin that is well suited for the remediation of a wide variety of environmental pollutants. This system includes mechanisms for detoxification that have proven effective for high concentrations of normally toxic compounds, as was demonstrated with trinitrotoluene and pentachlorophenol. Essential to successful detoxification of compounds is the presence of sufficient amounts of fungal mycelia. These systems normally provide protection from toxic products of lignin degradation. The wide variety of structurally unrelated chemicals degraded is due to the fact that the fungi have both oxidative and reductive non-specific mechanisms produced extracellularly. Mycoremediation can prove to be a proficient means of treating complex sites of contamination.

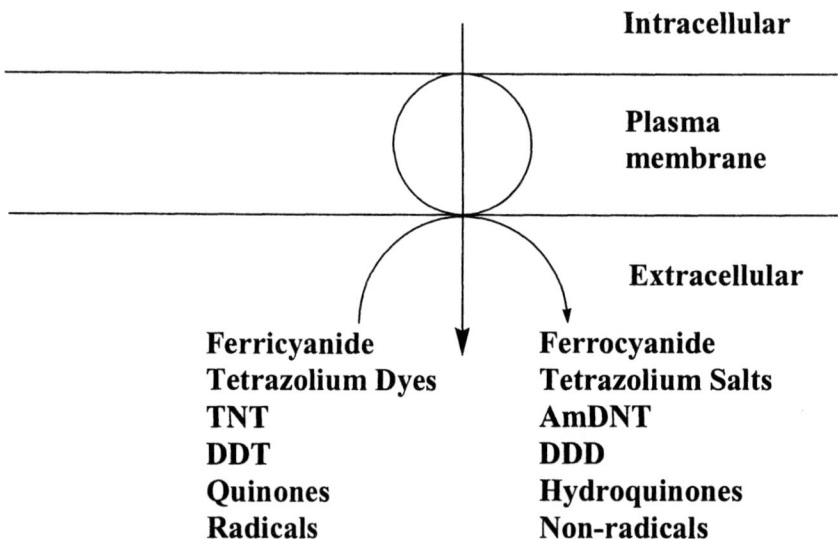

Figure 3. Chemicals reduced by fungal plasma membrane redox system.

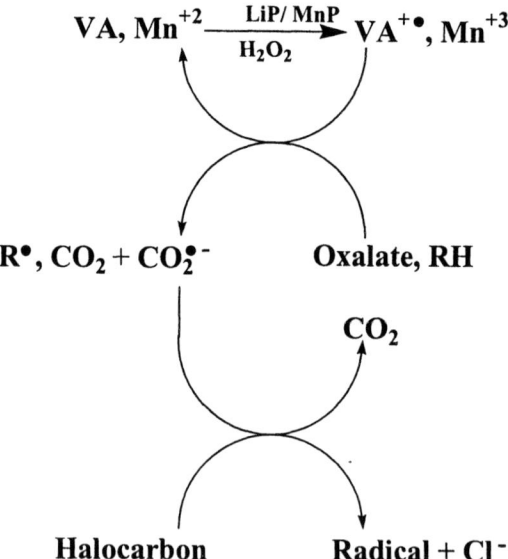

Figure 4 Reactions for aerobic dehalogenation of halocarbons by white-rot fungi.

References

1. Bumpus, J. A.; Tien, M.; Wright, D. S.; Aust, S. D. *Science* **1985**, 228, 1434-1436.
2. Camarero, S.; Barrasa, J. M.; Pelayo, M.; Martinez, A. T. *J. Pulp Paper Sci.* **1998**, 24, 197-203.
3. Eriksson, K. –E. L.; Blanchette, R. A.; Agner, P. In *Microbial and Enzymatic Degradation of Wood and Wood Components;* Eriksson, K. –E. L.; Blanchette, R. A.; Andra, P., Eds.; Springer-Verlag, Berlin, Germany, 1990, pp. 225-333.
4. Eichlerova, I.; Homolka, L.; Nerud, F.; Zadrazil, F.; Baldrian, P.; Gabriel, J. *Biodegradation* **2000**, 11, 279-287.
5. Otjen, L.; Blanchette, R.; Effland, M.;Leatham, G. *Holzforschung* **1987**, 41, 343-349.
6. Fernando, T.; Bumpus, J. A.; Aust, S. D. *Appl. Environ. Microbiol.* **1990**, 56, 1666-1671.
7. Hatakka, A. I.; Uusi-Rauva, A. K. *Eur. J. Appl. Microbiol. Biotechnol.* **1983**, 17, 235-242.
8. Gadd, G. M.; *Fungi in Bioremediation* Cambridge University Press, NY, 2001; p 481.
9. Stahl, J. D.; Aust, S. D. *Rev. in Toxicol.* **1998**, 2, 189-194.
10. Dean, J. F. D.; Eriksson, K. –E. L. *Holzforschung* **1992**, 46, 135-147.
11. Alder, E. *Wood Sci. Technol.* **1977**, 11, 169-218.
12. Lewis, N. G.; Yamamoto, E. *Ann. Rev. Plant Physiol. Plant Mol. Biol.* **1990**, 41, 455-496.
13. Fredenberg, K. In *Constitution and Biosynthesis of Lignin,* Neish, A. C.; Freudenberg, K., Eds.; Springer-Verlag, Ney York, 1986; p 45.
14. Crawford, R. L. *Lignin Degradation and Transformation;* Wiley; New York, NY, 1981.
15. Timofeevski, S. L.; Nie, G.; Reading, N. S.; Aust, S. D. *Arch. Biochem. Biophys.* **2000**, 373, 147-153.
16. Jenzer, H.; Jones, W,; Kohler, H. *J. Biol. Chem.* **1986**, 261, 15,550-15,556.
17. Khindaria, A.; Nie, G.; Aust, S. D. *Biochemistry* **1997**, 36, 14,181-14,185.
18. Kuan, I. –C.; Johnson, K. A.; Tien, M.; *J. Biol. Chem.* **1993**, 268, 20,064-20,070.
19. Wariishi, H.; Valli, K.; Gold, M.H. *Biochemistry* **1988**, 27, 5365-5370.
20. Shah, M. M.; Grover, T. A.; Barr, D. P.; Aust, S. D. *J. Biol. Chem.* **1992**, 267, 21564-21569.
21. Khindaria, A.; Grover, T. A.; Aust, S. D. *Environ. Sci. Technol.* **1995**, 29, 719-725.
22. Thurston, C. F.; *Microbiology* **1994**, 140, 19-26.
23. Bourbonnais, R.; Paice, M. G. *FEBS Lett.* **1990**, 267, 99-102.

24. Crosby, D. G. *Pure Appl. Chem.* **1981,** 53, 1051-1080.
25. Chung, N.; Aust, S. D. *J. Hazardous Materals,* **1995,** 41, 177-183.
26. Coulter, C.; Hamilton, J. T.G.; Harper, D. B. *Appl. Environ. Microbiol.* **1993,** 59, 1461-1466.
27. Harper, D. B.; Hamilton, J. T. G. *J. Gen. Microbiol.* **1988,** 134, 2831-2839.
28. Joshi, D. K.; Gold, M. H. *Appl. Environ. Microbiol.* **1993,** 59, 1779-1785.
29. Valli, K. Brock, B. J.; Joshi, D. K.; Gold, M. H. *Appl. Environ. Microbiol.* **1992,** 58, 221-228.
30. Schott, S.; Ruchoft, C. C.; Migregian, S. *Ind. Eng. Chem.* **1994,** 35, 1122-1132.
31. Fernando, T.; Aust, S. D. *ACS Symp. Ser.* **1991,** 468, 214-232.
32. Sollod, C. C.; Jennis, A. E.; Daub, M. E. *Appl. Environ. Microbiol.* **1992,** 58, 444-449.
33. Stahl, J. D.; Aust, S. D. *Biochem. Biophys. Res. Commun.* **1993,** 192, 477-482
34. Shah, M. M. ; Aust, S. D. *ACS Symp. Ser.* **1993,** 518, 191-202.
35. Stahl, J. D. ; Aust, S. D. In *Proceedings of the 9th Annual Conference on Hazardous Waste Remediation;* Kansas State Univ., Manhattan, 1994; pp 172-185.
36. Stahl, J. D. ; Aust, S. D. *Arch. Biochem. Biophys.***1995,** 320, 369-374.
37. Alexaner, M. *Environ. Sci. Technol.* **1985,** 19, 106-111.
38. Knecht, K. T.; DeGray, J. A.; Mason, R. P. *Mol. Pharmacol.* **1992,** 41, 943-949.
39. Khindaria, A.; Grover, T. A.; Aust, S. D. *Arch. Biochem. Biophys.* **1994,** 314, 301-306.

Chapter 2

Microbial and Photolytic Degradation of 3,5,6-Trichloro-2-pyridinol

Yucheng Feng

Department of Agronomy and Soils, Auburn University, Auburn, AL 36849

3,5,6-Trichloro-2-pyridinol (TCP) is a primary degradation product of the insecticide chlorpyrifos and the herbicide triclopyr. A bacterium, *Pseudomonas* sp. strain ATCC 700113, capable of using TCP as a sole source of carbon and energy, was isolated from a soil treated repeatedly with chlorpyrifos. TCP was metabolized to CO_2, chloride, and unidentified polar metabolites. *Pseudomonas* sp. ATCC 700113 immobilized on diatomaceous earth beads also mineralized [2,6-^{14}C]TCP rapidly; about 75% of the initial radioactivity was recovered as $^{14}CO_2$. Immobilized cells effectively removed TCP from wastewater generated from a chlorpyrifos-manufacturing plant; however, degradation of TCP was inhibited by high concentrations of NaCl. Photolysis of TCP occurred rapidly upon UV irradiation and released CO_2, chloride, dichlorodihydroxypyridine isomers, and reductive dechlorination products. Resting cell cultures of *Pseudomonas* sp. ATCC 700113 can only degrade the reductive dechlorination products in the mixture of photodegradation products, suggesting TCP degradation by this organism involves a reductive dechlorination pathway.

© 2004 American Chemical Society

Halogenated heterocyclic aromatic compounds are widely used for the production of pesticides, pharmaceuticals, and dyes, but much less is known regarding their metabolism by microorganisms compared with their homocyclic analogs. 3,5,6-Trichloro-2-pyridinol (TCP) is a major metabolite of the insecticide chlorpyrifos and the herbicide triclopyr. It has been detected in environments where chlorpyrifos and triclopyr were previously applied (1-6). TCP can be mineralized in soil, and its half-life varies with soil type, ranging from 10 to 325 days (5, 7). The mineralization of TCP in soil is microbially mediated, but isolation of the degradative microorganisms has rarely been attempted.

Chlorpyrifos and triclopyr readily degrade to TCP in the environment via photolysis and hydrolysis. Although relatively non-toxic to mammals, TCP exhibits low-to-moderate toxicity to aquatic and terrestrial biota (8). In addition, TCP has displayed some potential to affect microorganisms; soil concentrations higher than 100 ppm have been reported to retard the microbial degradation of several insecticides (5, 9, 10). In addition to the environmental matrices, TCP is also present in the raw, untreated wastewater originating from a chlorpyrifos-producing facility. Chemical oxidation is the conventional method to treat TCP-containing industrial wastewater. Biological methods (e.g., immobilized cell systems) are emerging as effective alternative treatment strategies, due to cost-effectiveness and lack of toxic by-product formation. Like its parent compounds, TCP undergoes photodecomposition, but little is known about the nature of the photodegradation products. Several researchers suggested that the combined use of photolysis and microbial metabolism might provide an alternative method for the treatment of chemical wastes (11-13).

The research presented in this paper deals with the isolation and characterization of a pure culture of TCP-degrading bacteria, treatment of TCP-containing wastewater using an immobilized cell system, photodegradation of TCP, and metabolism of TCP photolysis products by TCP-degrading bacteria.

Experimental Methods

Enrichment culture techniques were used to obtain TCP-degrading bacteria from an agricultural soil that had been treated repeatedly with chlorpyrifos (14). Identification of the TCP-degrading isolate was based on phenotypic characterization, substrate utilization pattern (Biolog, Inc.), and fatty acid profile analysis (MIDI, Inc.). The TCP-degrading isolate was grown in a mineral salt medium (14); 0.01% (w/v) yeast extract and 0.018% (w/v) glucose were added when large amounts of cells were needed. Mineralization of [2,6-^{14}C]TCP by the TCP-degrading bacterium was determined with growing, resting, and immobilized cells in batch cultures (14, 15). Diatomaceous earth pellets served as solid supports for immobilized cells. Glass columns (40 cm long and 4.9 cm in diameter) containing immobilized bacteria were used to evaluate the removal of TCP from industrial wastewater (15).

For photodegradation experiments, TCP solutions were irradiated with an ultraviolet (UV) light source (254 nm) at room temperature for various periods of time depending on the purpose of each experiment (16). The photodegradation products were extracted with ethyl acetate and analyzed by thin-layer chromatography (TLC) and gas chromatography-mass spectrometry (GC-MS). The samples derivatized with bis(trimethylsilyl)trifluoroacetamide (BSTFA) were also analyzed by GC-MS. [2,6-^{14}C]TCP was used in some experiments to monitor $^{14}CO_2$ evolution and product formation. Unknown products with high radioactivity on TLC plates were isolated and characterized using mass spectrometry and proton nuclear magnetic resonance analyses. Metabolism of TCP photolysis products by resting cells of TCP-degrading bacteria was monitored by analyzing ethyl acetate extracts of the reaction solution using GC (16).

Results and Discussion

Isolation and characterization of TCP-degrading bacteria

A bacterium capable of using TCP as a sole source of carbon and energy was isolated from a soil with previous exposure to chlorpyrifos, using enrichment culture techniques. The organism was identified as a *Pseudomonas* sp., deposited in the general collection of the American Type Culture Collection, and assigned the accession number ATCC 700113. *Pseudomonas* sp. ATCC700113 was a Gram-negative rod; colonies formed on agar plates were circular with an entire margin and smooth surface. Colonies appeared yellow on nutrient agar and yeast-extract/glucose agar, but a brown pigment was produced on tryptic soy and King's B agar media. The pigment was non-fluorescent and diffusible on King's B plates. The isolate was catalase positive, cytochrome oxidase positive, and arginine dihydrolase negative. It reduced nitrate to nitrite, used glucose oxidatively, and hydrolyzed gelatin but not starch. Cells grown on TCP did not cleave the aromatic ring of catechol at either *ortho* or *meta* position according to the method described by Smibert and Krieg (17).

Pseudomonas sp. ATCC 700113 was able to mineralize [2,6-^{14}C]TCP readily. About 72.4% of the initial radioactivity was recovered as $^{14}CO_2$, 9% remained in the medium, and 4.1% in biomass. Chloride was released stoichiometrically and concurrently with the disappearance of TCP. Manipulation of experimental conditions revealed that TCP was most rapidly degraded at a temperature of 28°C and pH in the 6.2 to 7.8 range. TCP at concentrations of 100 mg/L or less was degraded rapidly, but at 200 mg/L no transformation occurred, perhaps due to its toxic effect on the bacteria. Transformation of TCP was also affected by the presence of NaCl; degradation rates decreased with increasing concentrations of NaCl and degradation was completely inhibited at a NaCl concentration of 10 g/L. Supplementation with a second carbon source such as glucose, maleic acid, and succinic acid was shown to stimulate bacterial growth as well as TCP degradation. This suggests that cells were carbon starved while growing on TCP alone.

Resting cells of *Pseudomonas* sp. ATCC 700113 metabolized TCP (40 mg/L) within 70 hours at 28°C. About 87% of the initial radioactivity was recovered as $^{14}CO_2$, 3% remained in the medium, and 1.6% was associated with the biomass (Figure 1). HPLC chromatograms revealed the presence of some polar metabolites; however, the amounts were insufficient for further characterization. Resting cells, however, did not metabolize 2-, 3-, and 4-hydroxypyridines, 2,4,5-trichlorophenol, and 2,4,5-trichloroaniline. Currently, *Pseudomonas* sp. ATCC700113 is the only known pure culture of bacteria capable of mineralizing TCP and using it as a sole source of carbon and energy.

Treatment of TCP-containing industrial wastewater

Immobilized bacteria have the potential to degrade chemical wastes faster than conventional wastewater treatment systems since high densities of bacteria are used in immobilized cell systems (18-20). Diatomaceous earth pellets are good support materials for the immobilization of bacteria due to their large surface area, high thermal and chemical stability, and mechanical strength. *Pseudomonas* sp. ATCC 700113 immobilized on diatomaceous earth pellets degraded [2,6-^{14}C]TCP (40 mg/L) in seven days. At the end of the experiment, 75.6% of initial radioactivity was recovered in the form of $^{14}CO_2$ and 2.5% of initial radioactivity remained in the medium that can be attributed to the polar metabolites formed during the degradation process. Chemical oxidation is the most common method used to remove TCP from industrial wastewater. Numerous organic by-products form during the chemical treatment process. Results from this study indicate that the metabolites remaining after microbial degradation of TCP were negligible.

Using columns containing immobilized bacteria, TCP removal of 80 to 100% was achieved with gradual increases of flow rates (from 0.12 to 0.73 ml/min) and TCP concentrations (from 60 to 140 mg/L) in the influent. The column inoculated with growing cells showed immediate removal of TCP after the inoculation while there was a lag phase for the column inoculated with resting cells. Figure 2 indicates that 100% removal of TCP (70 mg/L) was achieved when salt concentrations remained at 0.3%; however, reduced TCP

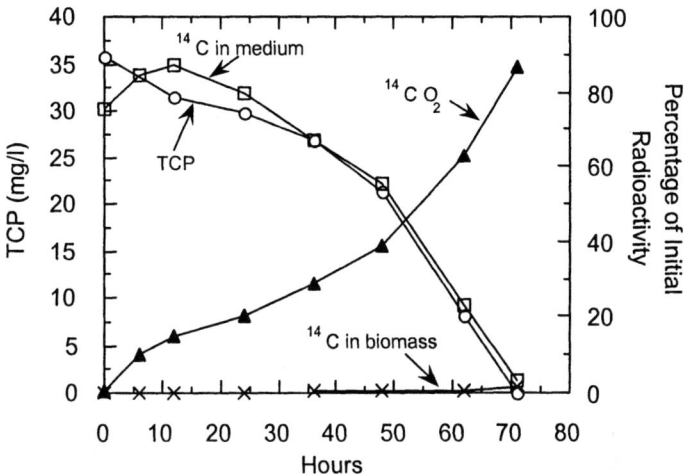

Figure 1. Metabolism of TCP by resting cells of Pseudomonas sp. ATCC 700113. (Reproduced with permission from reference 14. Copyright 1997 American Society for Microbiology.)

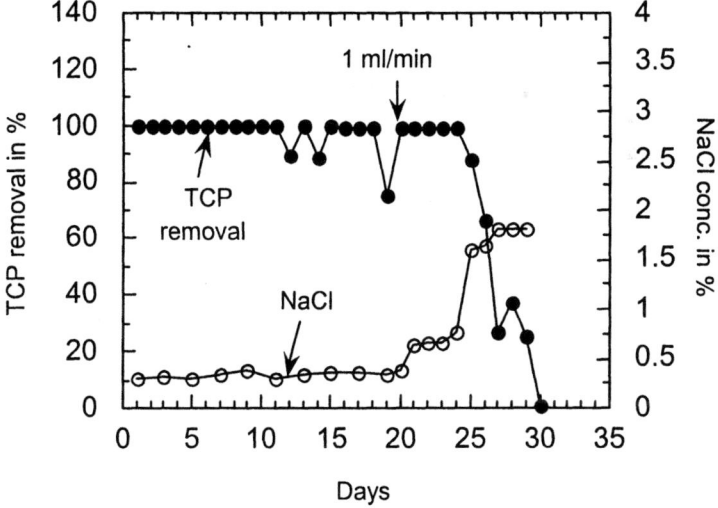

Figure 2. Effect of sodium chloride concentration on the removal of TCP by immobilized Pseudomonas ATCC 700113. The flow rates were gradually increased to 1mL/min. (Reproduced with permission from reference 17. Copyright 1997 Springer-Verlag.)

removal was observed when sodium chloride concentration was increased. TCP removal was completely inhibited when sodium chloride concentration reached 1.8%. These results showed that *Pseudomonas* sp. ATCC 700113 immobilized on diatomaceous earth pellets effectively removed TCP from wastewater at low salt concentrations. Growing cell cultures were found to be better inocula for immobilization than resting cells. Since large amounts of salts are present in wastewater generated from chlorpyrifos-manufacturing plants, the tolerance of microorganisms to salts is important for a microbial treatment process. *Pseudomonas* sp. ATCC 700113 cannot tolerate a sodium chloride concentration of 10 g/L and addition of 1 mM osmoprotectants (i.e., betaine and proline) did not reduce the adverse effect of salt. Further research is needed to develop TCP-degrading bacteria that tolerate high salt concentrations.

Photodegradation of TCP and biodegradation of TCP photolysis products

TCP has been shown to undergo rapid photodegradation in aqueous solutions and on natural and synthetic surfaces (8). Although the complete pathway of TCP photolysis has not been established, photodehalogenation and CO_2 production have been observed (21). Upon exposure to UV light, TCP (80 mg/L) disappeared to nondetectable levels within 2 hours and the amount of chloride released was 85% of the calculated stoichiometric amount (Figure 3). The color of the reaction solution changed from colorless to reddish brown. Approximately 25% of the initial radioactivity was recovered as $^{14}CO_2$, and 67% remained in the reaction solution (16). Several products resulting from photodegradation of TCP were identified; they are isomers of dichlorodihydroxypyridines (**I**), isomers of chlorodihydro-2-pyridone (**II**), tetrahydro-2-pyridone (**III**), maleamide semialdehyde (**IV**), and isomers of dichlorocyanopropene. Formation of these products suggests that both reductive dechlorination and hydrolytic dechlorination mechanisms were involved in TCP photolysis. A proposed photodegradation pathway is shown in Figure 4.

Mixtures of photodegradation products were used as substrates for resting cells of *Pseudomonas* sp. ATCC 700113. After 4-day incubation, most TCP photodegradation products were degraded, with the exception of dichlorodihydroxypyridines. This observation suggests that *Pseudomonas* sp. ATCC 700113 used the reductive dechlorination pathway to degrade TCP. This finding could also explain why *Pseudomonas* sp. ATCC 700113 gave a negative result for the aromatic ring cleavage test.

Although TCP was degraded rapidly upon UV irradiation, the extent of mineralization was much less compared to microbial degradation. A combined use of photolysis and microbial metabolism has been suggested as an alternative

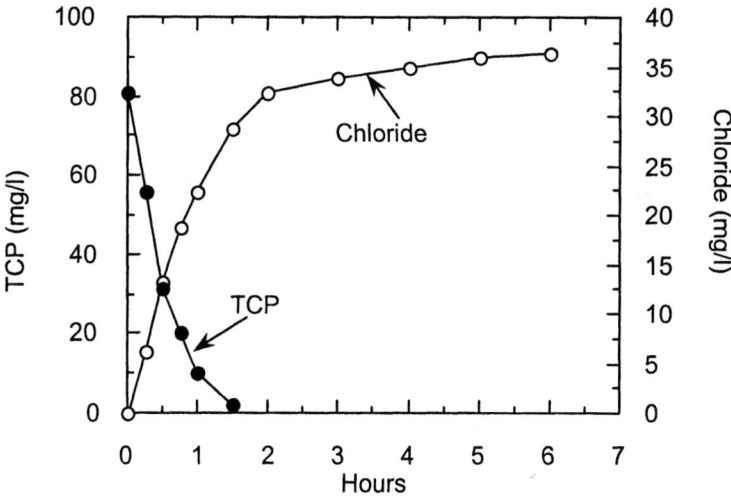

Figure 3. Photodegradation of 3,5,6-trichloro-2-pyridinol upon exposure to UV light. (Reproduced with permission from reference 18. Copyright 1998 Society of Environmental Toxicology and Chemistry.)

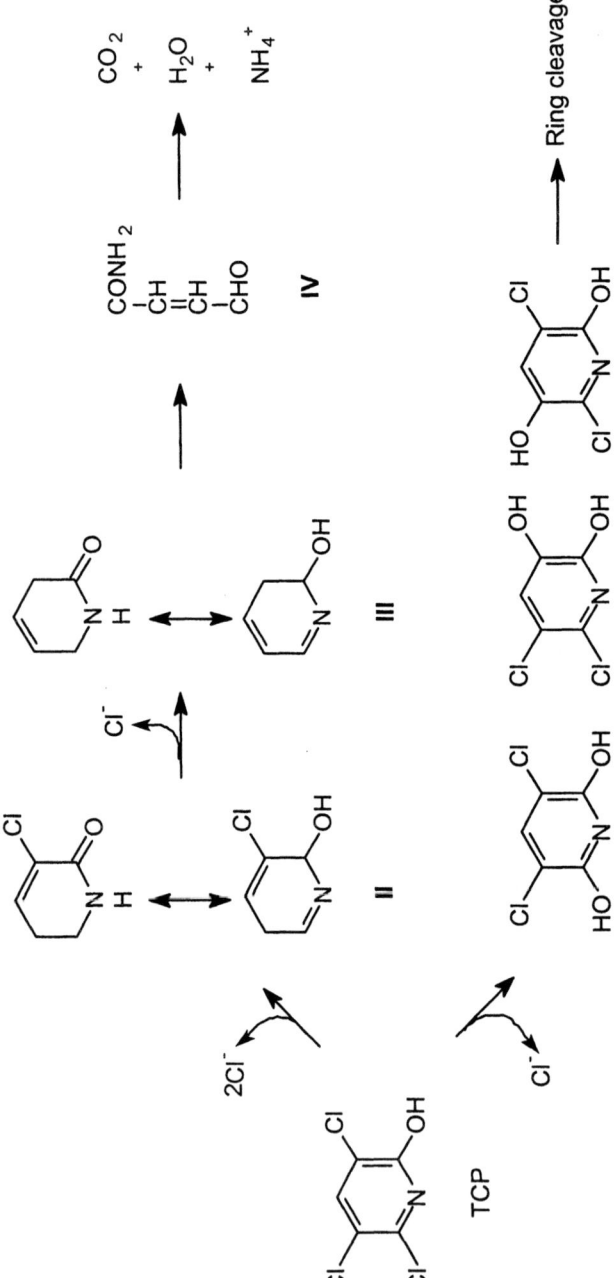

Figure 4. Proposed photodegradation pathway of 3,5,6-trichloro-2-pyridinol. (Reproduced with permission from reference 18. Copyright 1998 Society of Environmental Toxicology and Chemistry.)

method for the treatment of chemical wastes (11-13). Due to the free radical nature of photolysis, products of the reaction can be complex. A consortium of microorganisms may be needed to achieve a complete degradation in the treatment process combining microbial and photolytic activity.

Conclusions

Pseudomonas sp. ATCC 700113, capable of using TCP as a sole source of carbon and energy, was isolated from a soil with previous exposure to chlorpyrifos. This organism appeared to utilize a reductive dechlorination mechanism to degrade TCP and produce CO_2, chloride, and water as end products. Immobilized cells of *Pseudomonas* sp. ATCC 700113 effectively removed TCP from industrial wastewater at low salt concentrations. The salt inhibition effect must be overcome before a biological treatment method can be implemented in treating wastewater generated from chlorpyrifos-manufacturing plants.

References

1. Chapman, R. A.; Harris, C. R. *J. Environ. Sci. Hlth.* **1980**, *B15*, 39-46.
2. Getzin, L. W. *J. Econ. Entomol.* **1981**, *74*, 158-162.
3. Lee, C. H.; Oloffs, P. C.; Szeto, S. Y. *J. Agric. Food Chem.* **1986**, *34*, 1075-1079.
4. Norris, L. A.; Montgomery, M. L.; Warren, L. E. *Bull. Environ. Contamin. Toxicol.* **1987**, *39*, 134-141.
5. Racke, K. D.; Coats, J. R.; Titus, K. R. *J. Environ. Sci. Hlth.* **1988**, *B23*, 527-539.
6. Thompson, D. G.; Staznik, B.; Fontaine, D. D.; Mackay, T.; Oliver, G. R.; Troth, J. *Environ. Toxicol. Chem.* **1991**, *10*, 619-632.
7. Racke, K. D.; Robbins, S. T. In *Pesticide transformation products: fate and significance in the environment*; Somasundaram, L., Coats, J. R., Eds.; American Chemical Society: Washington, DC, 1991; pp 93-107.
8. Racke, K. D. *Rev. Environ. Contamin. Toxicol.* **1993**, *131*, 1-154.
9. Racke, K. D.; Fontaine, D. D.; Yoder, R. N.; Miller, J. R. *Pesticide Sci.* **1994**, *42*, 43-51.
10. Chapman, R. A.; Harris, C. R. In *Enhanced biodegradation of pesticides in the environment*; Racke, K. D., Coats, J. R., Eds.; American Chemical Society: Washington, DC, 1990; pp 82-96.
11. Acher, A. J.; Hapeman, C. J.; Shelton, D. R.; Muldoon, M. T.; Lusby, W. R.; Avni, A.; Waters, R. *J. Agric. Food Chem.* **1994**, *42*, 2040-2047.

12. Amador, J. A.; Taylor, B. F. *Appl. Environ. Microbiol.* **1990**, *56*, 1352-1356.
13. Somich, C. J.; Kearney, P. C.; Muldoon, M. T.; Elsasser, S. *J. Agric. Food Chem.* **1988**, *36*, 1322-1326.
14. Feng, Y.; Racke, K. D.; Bollag, J.-M. *Appl. Environ. Microbiol.* **1997**, *63*, *4096-4098*.
15. Feng, Y.; Racke, K. D.; Bollag, J.-M. *Appl. Microbiol. Biotechnol.* **1997**, *47*, 73-77.
16. Feng, Y. *Environ. Toxicol. Chem.* **1998**, *17*, 814-819.
17. Smibert, R. M.; Krieg, N. R. In *Methods for general and molecular bacteriology*; Krieg, N. R., Ed.; American Society for Microbiology: Washington, DC, 1994; pp 607-654.
18. Hallas, L.; Adams, W. J.; Heitkamp, M. A. *Appl. Environ. Microbiol.* **1992**, *58*, 1215-1219.
19. Heitkamp, M. A.; Adams, W. J. *Can. J. Microbiol.* **1992**, *38*, 921-928.
20. Rothenburger, S.; Atlas, R. M. *Appl. Environ. Microbiol.* **1993**, *59*, 2139-2144.
21. Smith, G. N. *J. Econ. Entomol.* **1968**, *61*, 793-799.

Chapter 3

Biological Detoxification of Organophosphate Pesticides

Mark Shimazu[1], Wilfred Chen[1,2], and Ashok Mulchandani[1,2,*]

[1]Department of Chemical and Environmental Engineering and
[2]Environmental Toxicology Program, University of California, Riverside, CA 92521

> The diverse array of naturally isolated microorganisms capable of degrading organophosphate pesticides has led to increased attention in using bioremediation as a method to treat these compounds. Combined with advancements in biotechnology, manipulation of natural systems for novel strategies in pesticide degradation has provided an environmentally-friendly in situ treatment method. The innovative use of bifunctional proteins with a catalytic domain and moieties for cell surface targeting, facilitated purification and immobilization are some of the recent strategies used to treat these chemical agents. This chapter will focus on the use of genetically engineered microorganisms and custom tailored enzymes as an alternative to current disposal methods for the treatment of organophosphate pesticides.

Introduction

To feed the growing population of the world, modern agriculture has relied heavily on pesticides to produce a high yield of crops. Although the use of pesticides has increased crop production by many-fold, the use of pesticides is not without its consequences. The efficacy of pesticides is due to their ability to act as a neurotoxin to the pests they control. Unfortunately pesticides are not only harmful to their intended targets, but to all animals, including humans.

One of the most used pesticides throughout the world is the organophosphate (OP) class of compounds. OP compounds are very effective, widely available and relatively inexpensive making them attractive to both developed as well as developing nations. In the United States alone over 40 million kg of organophosphate pesticides are applied annually (*1*). OP compounds are available for industrial, agricultural and home use. Commonly used OP pesticides include parathion, coumaphos, dursban, diazinon and malathion. Although recent studies on the environmental impact of OP pesticides have led to the reduced use or outright banning of OP compounds in more industrialized nations, they are still used extensively in poorer and developing nations.

Organophosphates and their family of compounds are potent neurotoxins that share structural similarities to chemical warfare agents such as sarin, soman and VX. Organophosphates are cholinesterase inhibitors and exposure to OP compounds leads to increased neurotransmitter accumulation. Classical symptoms of OP exposure include salivation, lacrimination, urination and defecation. Exposure to OP compounds can cause fatigue, dizziness, vomiting, paralysis and even death (*2*).

As a result of their toxicity and wide spread usage, there is a need to treat the large amounts of wastes from unused pesticide concentrates, agricultural runoff, accidental spillage, and residue from cleaning of spray equipment and storage tanks. Clearly there is a need for the rapid and efficient methods by which to dispose of these compounds.

Conventional disposal of OP compounds involves chemical treatment (hydrolysis), incineration and placement in landfills. The threat of leaching and groundwater contamination makes the use of landfills very undesirable and chemical hydrolysis relies on the use of harsh acids and bases whose byproduct's be disposed of and treated. Incineration remains the only EPA approved method for OP disposal but is publicly opposed due to the potential release of harmful and toxic byproducts into the atmosphere (*3,4*). Due to the fact that current disposal methods inadequately address the removal of OP compounds, new more environmentally friendly methods need to be developed (*5,6*).

Microbial Degradation

Naturally occurring bacteria isolates capable of metabolizing organophosphates have received considerable attention since they provide the possibility of both environmentally-friendly and in-situ detoxification. Soil microorganisms *Pseudomonas diminuta* MG and *Flavobacterium* sp. have been shown to express high activities of organophosphorus hydrolase (OPH), which hydrolyzes organophosphate pesticides effectively detoxifying them. Hydrolysis of parathion reduces its toxicity by nearly 120-fold and leads to the formation of *p*-nitrophenol (7).

Pseudomonas diminuta MG and *Flavobacterium* sp., have been well characterized and the genes (opd) coding for hydrolase activity has been cloned and sequenced. The two genes were also shown to be identical on the amino acid level (8). The hydrolases from both bacteria have also been purified and extensive kinetic characterizations have been preformed (9,10).

Table I. OPH Kinetic Properties

Compound	$K_{cat}(s^{-1})$	K_m (mM)	K_{cat}/K_m ($M^{-1}s^{-1}$)
Paraoxon	3180	0.058	5.5×10^7
Parathion	630	0.24	2.6×10^6
Methyl parathion	189	0.08	2.4×10^4
Diazinon	176	0.45	3.9×10^6
DFP	465	0.048	9.7×10^4
Sarin	56	0.7	80×10^3
Soman	5	0.5	10×10^3
Demeton-S	1.3	0.78	1.6×10^3

Organophosphorus hydrolase (OPH, EC 3.1.8.1) is a homodimer with a binuclear metal center. OPH has broad substrate specificity and can hydrolyze organophosphate pesticides such as methyl parathion, ethyl parathion, paraoxon, chlorpyrifos, coumaphos, cyanophos and diazinon. Table I (9,12-14). The enzymatic hydrolysis rates are 40 – 2450 times faster than chemical hydrolysis at pH 7.0 and the enzyme is reported to be stable at temperatures of up to 45-50°C (3). However, hydrolysis rates varied from very fast for phosphotriesters and phosphothiolester pesticides (P-O bond) such as paraoxon ($k_{cat} > 3800s^{-1}$) and coumaphos ($k_{cat} = 800s^{-1}$) to limited hydrolysis for Diazinon ($k_{cat} = 176 \ s^{-1}$) and fensulfothion ($k_{cat} = 67 \ s^{-1}$) (14).

The opd gene has been expressed in a wide variety of heterogeneous hosts including *E. coli* (15), insect cell (fall armyworm) (16), *Streptomyces* (17), and soil fungus (18). *E. coli* has been the most widely used and studied in large part due to ease of growth and maintenance and the high cell densities achieved during fermentation. Initial attempts at expression using the native promoter resulted in very poor expression levels (*19,20*). However, increased levels of expression were achieved with the use of a strong *lac* promoter (*10*).

Enzyme Detoxification of OP Neurotoxins

OPH Immobilization by Affinity Tags

Physical adsorption and covalent attachment of both native and recombinant OPH onto various supports such as nylon membranes, porous glass and nanometer size silica beads have been employed (*3,21,22*).. Unfortunately, physical adsorption offers poor and nonspecific binding, while covalent modifications to OPH often results in reduction of enzyme activity and kinetic properties (*22,23*). In addition to reducing catalytic activity, there is no controlled orientation of the immobilized enzymes, leading to inaccessibility of the substrate to the enzyme active site. In the case of covalent bonding, the immobilization support is not reusable since the formed covalent bond is irreversible. In addition, the tedious and costly protocol for purification of OPH limits its use in large-scale enzymatic degradation.

An alternative to physical adsorption and covalent cross-linking is the use of an affinity tag fused to the OPH moiety enabling one-step purification and oriented immobilization. Affinity tags offer strong reversible binding under mild non-denaturing conditions as well as proper orientation of the enzyme for full substrate accessibility to the enzyme active site.

Recently, a fusion protein between a cellulose binding domain (CBD_{clos}), isolated from *Clostridium cellulovorans*, and OPH was shown to be capable of immobilization onto various cellulose materials (*24*). The use of cellulose as an immobilization matrix is advantageous due to its low cost, wide spread availability and non-toxic nature. The kinetic parameters of OPH fused to the CBD domain were essentially identical to the soluble protein with a 3.8 fold increase in K_m from 0.058 to 0.220 and a 10.4% decrease in K_{cat} from 3170 to 2840. Additionally, the immobilized fusion protein offered superior stability over that of soluble OPH, retaining over 85% activity over a period of 45 days (*24*).

In another strategy to orient OPH correctly onto a solid support, an octapeptide, Asp-Tyr-Lys-Asp-Asp-Asp-Asp-Lys (FLAG), was fused to organophosphorus hydrolase. The OPH-FLAG fusion protein was then immobilized onto magnetic beads coated with a protein A–anti-FLAG monoclonal antibody. The fusion of the OPH to the FLAG showed no detrimental affect on enzyme activity. Free soluble OPH was measured with a catalytic efficiency (K_{cat}/K_m) of 2.9 ($\pm$0.3) x $10^7 M^{-1} s^{-1}$ while the fusion exhibited similar efficiency of 2.5 ($\pm$0.3) x $10^7 M^{-1} s^{-1}$. However, when immobilized onto the magnetic beads, there was a 2-fold increase in K_m and a 90% decrease in K_{cat}. Stability of the enzyme improved with a reported half-life of 23 days when immobilized and stored at 4°C (25).

A bifunctional fusion protein consisting of organophosphorus hydrolase (OPH) and elastin-like polypeptide (ELP) has also been synthesized for the detoxification of organophosphorus compounds. Elastin-like proteins (ELPs) are composed of repeats of the pentapeptide sequence Val-Pro-Gly-Xaa-Gly (Where Xaa is any residue except for proline) and undergoes a sharp temperature transition from soluble to insoluble with an increase of ionic strength of the buffer or increase in temperature (26). The phase transition is a hydrophobically driven process causing the side chains of hydrophobic amino acid residues aggregating to a thermodynamically favorable state with the increase in temperature or ionic strength (27). This thermally triggered property of phase transition allowed for a simple and rapid means of purifying the fusion protein. Over 1300-fold purification was achieved after only 2 cycles of inverse phase transition. The purified fusion protein showed essentially identical kinetic properties as native OPH with only a 10% increase in K_m and a 5% increase of K_{cat}. The ability of the ELP domain to form collapsed aggregates also improved long-term stability of the fusion enzyme. Aggregated ELP-OPH retained nearly 100% activity over a span of three weeks. In addition to facilitating purification and stability, the ELP moiety served as a hydrophobic tag for one-step immobilization of the fusion protein onto hydrophobic surfaces. The ELP-OPH was capable of rapidly degrading paraoxon while immobilized onto the surface. The protein also retained ELP functionality of reversible phase transition thereby allowing for the regeneration of the treated surface (28).

Whole Cell Detoxification of OP Neurotoxins

A major economic obstacle in using enzymes to degrade OP compounds is the need to purify large amounts of enzyme. Traditionally, gel filtration and ion exchange chromatography are used. However these methods are costly, tedious and are not amicable to scale up. A strain of *Streptomyces lividans* has been developed for the secretion of OPH into the medium (17). This method allows for minimal cell wall disruption and handling of the cells, but still required additional purification of the end product. To alleviate the need for any steps in purifying protein, whole cells were used as live biocatalyst instead of purified proteins.

The use of whole cells as live biocatalysts for the detoxification of OP compounds has been demonstrated using *Pseudomonas putida* (*29*) and *Pseudomonas* sp. A3 (*30*). *P. putida* immobilized using 3% sodium alginate beads (w/v) and a wet cell mass of 5% (w/v) was capable of degrading 99% of 1mM methyl parathion during the course of a 48 hour incubation time. Further, the reactor was re-usable for numerous batch runs (*30*).

A microbial consortium was isolated from soil contaminated with coumaphos and used in a biofilter. The consortium was colonized onto sand, gravel and Celite and packed into a bio-trickling filter. After 7-10 days at 25°C, the coumaphos concentrations dropped from 1200mg/L to between 0.010 and 0.050 mg/L (*31*). Subsequently, a field-scale bioreactor capable of treating 15,000 L batches was constructed. In large scale runs, the bioreactor was capable of reducing the coumaphos concentrations from 2000mg/L to 10mg/ml. The reactor was capable of a 200-fold reduction in coumaphos in two successive 11,000L batch runs. However, fouling became a problem with prolonged use (*32*).

Surface Expression of OPH

When expressed intracellularly, the cell membrane can limit substrate access to the enzyme. In effect the cell membrane acts as a diffusional barrier whereby the enayme cannot freely access the substrate. It has been previously shown that substrate uptake was the rate-limiting step in degradation of these pesticides by recombinant *E. coli* expressing OPH intracellularly (*24,33,34*). By expressing the OPH on the surface of the cell all diffusional limitations caused by the cell membrane can be alleviated.

Using the Lpp-OmpA anchor system, an Lpp-OmpA-OPH fusion was targeted successfully onto the surface of *E. coli*. The fusion was targeted with over 90% efficiency and afforded a seven-fold increase in hydrolysis of parathion when compared to cells expressing similar amounts of OPH intercellularly. In addition the increased efficiency, the live biocatalysts were more stable than soluble OPH and remained active over the span of 30 days (*24*).

More recently, OPH was targeted onto the surface of an *E. coli* and *Moraxella* sp. using an ice-nucleation protein (INP) anchor derived from *Pseudomonas syringe* (*35,36*). Ice-nucleation protein (INP), an outer membrane protein from *Pseudomonas syringae*, is composed of three distinct domains: an N-terminal domain (15%) containing three or four potential transmembrane spans, a C-terminal domain (4%), and a central domain composed of repeating residues which act as the template for ice nucleation (*37*). The INP portion of the INP-OPH fusion was truncated and the central repeating domain was deleted because it was not required for targeting the protein to the cell surface (*36*). Combined with the native *p*-nitrophenol degrading pathway, the genetically engineered *Moraxella* sp. with surface expressed OPH was capable of completely degrading 0.4 mM paraoxon and it's hydrolysis product within 10 hours. The resulting *Moraxella* sp. degraded organophosphates as well as PNP rapidly, all within 10 h. The initial hydrolysis rate was 0.6 μmol/h/mg dry weight, 1.5 μmol/h/mg dry weight, and 9.0 μmol/h/mg dry weight for methyl parathion, parathion, and paraoxon, respectively. More impressively, *Moraxella* sp. was capable of an 80-fold increase in OPH activity relative to *E. coli* (*36*) providing a potentially powerful platform for microbial degradation of various toxic compounds using this robust bacterium.

Cells with surface-expressed OPH were immobilized on a non-woven polypropylene fabric support and effectively degraded 90-100% of diazinon, methyl parathion, paraoxon, and coumaphos (*1*) in less than 3 h. The ability of immobilized cells to carryout repeated cycles of degradation in sequencing batch reactor was demonstrated and only a small decline in degradation performance was detected during 12 consecutive repeated sequence batch degradation of paraoxon over a period of 19 days. In addition to polypropylene supports, cells were also immobilized onto Siran^{TM} porous glass beads. The reactor degraded 100% of 0.2 mM coumaphos in 2 hours. The flow rate was 23mL/h (*38*).

A genetically engineered *Escherichia coli* cell expressing both organophosphorus hydrolase (OPH) and a cellulose-binding domain (CBD) on the cell surface was constructed, enabling the simultaneous hydrolysis of organophosphate nerve agents and immobilization via specific adsorption to cellulose. With surface expressed CBD on the surface of bacteria, loss of whole cell bioreactor activity from gradual cell detachment was overcome. OPH was displayed on the cell surface by use of the truncated ice nucleation protein (INPNC) fusion system, while the CBD was surface anchored by the Lpp-OmpA fusion system. The surface anchored CBD immobilized 250mg/cm^2 (dry weight) of cells forming a monolayer as observed under an electron microscope. Combined with surface expressed OPH, the coexpression system was capable of degrading paraoxon very rapidly with the initial rate of 0.65mM/min/g (dry weight) of cell. The bioreactor also retained nearly 100% efficiency over the period of 45 days (*39*).

Modifications of Specificity and Activity

Site-directed Mutagenesis

Although OPH hydrolyzes a wide range of organophosphates, the effectiveness of hydrolysis varies dramatically. For example, some widely used organophosphorus insecticides such as methyl parathion, chlorpyrifos, and diazinon are hydrolyzed 30 to 1,000 times slower than the preferred substrate, paraoxon [9]. This reduction in catalytic rate is due to the unfavorable interaction of these substrates with the active sites involved in catalysis and/or structural functions (40). Although OP degradation remains an attractive technology, catalytic efficiencies for a number of OP compounds leave room for improvement.

The crystal structure of OPH was elucidated in 1995 (41). OPH is composed of a distorted α/β barrel with eight parallel β –pleated sheets making up the barrel structure linked to the outer surface by 14 α -helices (41,42). The active site contains a divalent metal complexed through three histidine residues with preference for Co^{+2} as the catalytic center (9,43). Site directed mutagenesis of key residues was attempted to improve overall catalytic efficiency of the enzyme.

Initial efforts focused on the substitution of amino acids directly responsible for binding the divalent metals. Site directed mutagenesis was used to substitute the original histidine residues at positions 254 and 257 (44,45). Three of the variants (H254R, H254S and H257L) all resulted in the enzyme containing only one of the original two metals at the bimetal nuclear center. Enzyme activity towards the nerve agent VX (O-ethyl-S-(2-diisopropyl aminoethyl) methylphosphonothioate) and insecticide analog demeton-S was improved four to five-fold while retaining the ability to hydrolyze paraoxon. The double mutant of H254R/H257 had a 20-fold increase activity towards demeton-S over wild type (44,45).

Residues surrounding the leaving group pocket were also substituted. Targeted here was the small substrate, DFP (diisopropyl fluorophosphate- a sarin analog). Hydrophobic amino acids (Trp131, Phe132, Leu271, Phe306, and Try309) were used to substitute those residues lining the leaving group. Because DFP has a fluoride leaving group, replacement of one side chain with a residue capable of hydrogen bond formation and proton donation (His, Tyr, or Lys) was predicted to enhance catalysis. K_{cat} of Phe132 and Phe306 and some double mutants showed a tenfold increase (46). Design of mutagenesis experiments using information derived from the crystal structure of OPH and its homology with acetylcholinesterase were attempted. In this rational design approach, a L136Y mutant showed a 33% increase in the relative VX hydrolysis rate compared to wild type enzyme (47).

Directed Evolution

DNA shuffling has received considerable attention in the past decade. In contrast to rational design procedures, mutation, selection and recombination occur in nature resulting in evolution of highly adapted enzymes from the infinite possibilities encoded in the genome. Recent advance in molecular techniques has now provided the possibility to mimic these natural evolutionary processes. Unlike natural selection, in which multiple environmental forces select enzymes to meet a variety of challenges, directed evolution exerts focused selection, enabling the rapid development of variants with highly specialized traits (*48*). In such cases, random mutagenesis combined with focused selection is a useful alternative for generating both the desired improvements and a database for future rational approaches to protein design. Screening of new desirable features can take place under progressively selective conditions.

DNA shuffling (*48,49*) is a method for *in vitro* homologous recombination of pools of selected genes by random fragmentation and PCR reassembly. It involves digesting a large gene with DNase I to form a pool of random DNA fragments. These fragments can be reassembled into a full-length gene by repeated PCR cycles. The fragments prime each other based on homology, and recombination occurs when fragments from one copy of a gene prime another copy, causing a template switch. In this fashion, positive mutants can be easily shuffled and recombined. An added benefit is the possibility to accelerate the rate of directed evolution by generating additional mutations even during the recombination and shuffling process. Moreover, nonfunctional mutations can be eliminated by backcross cycles (*50*). DNA shuffling combined with focused selection pressure has been applied to evolve genes for a variety of applications. This approach has been applied successfully to enhance or alter various enzyme features, including thermal stability (*50*), stability in organic solvent (*51*), activity (*48,52*) and substrate specificity (*53,54*). The technique also allows for the generation of mutants distal to the active site.

Recently this technique was applied to the evolution of OPH for increased methyl parathion hydrolysis. After two rounds of DNA shuffling, a variant that could hydrolyze methyl parathion 25-fold faster than wild type was isolated. The mutations were not directly located in the active site and could not be otherwise predicted *a priori* (*55*). This technique could be used to target other slow degrading pesticides such as chlorpyrifos and diazinon and against chemical warfare agents VX and sarin.

Conclusion

The use of pesticides in agriculture is inevitable and necessary. The high crop yields needed to feed the world's growing population would not be attainable without them. Unfortunately pesticides are not a silver bullet solution to pest control. Although very effective, they produce unwanted toxicity in humans and other nontarget organisms.

Microbial and enzymatic approaches towards the treatment of pesticides have received substantial attention in the past decade. Using current molecular biology approaches allows for an efficient and environmentally friendly method by which to degrade toxins. The ability to target enzymes onto the surface of cells is a powerful strategy in overcoming the limitations of using live biocatalysts in the treatment of OP compounds. Affinity tag purification has alleviated the tedious protocols of traditional column purification. The CBD-OPH and ELP-OPH fusion system offers an attractive one-step purification and immobilization of enzymes. The advent of DNA shuffling and directed evolution strategies has opened up an entire new realm in protein engineering where an enzyme can be "tuned" to target a wide array of substrates. Although the use of these technologies in large-scale operations is limited to date, we will eventually turn to these methods as the technology matures and as the public starts to demand the use of more environmentally friendly treatment methods.

References

1. Mulchandani, A.; Kaneva, I.; Chen, W. *Biotechnol. Bioeng.* 1999, 63, 216-223.
2. Casarett and Doull's Toxicology: the basic science of poisons. Klaassen, C.D., Amdur, M.O., Doull J., 5th ed. McGraw-Hill: New York, NY, 1996.
3. Munnecke, D.M. *Biotechnol. Bioeng.* 1979, 21, 2247-2261.
4. Munnecke, D.M.; Day, R.; Trask, H.W. *Rev. Pest. Disposal Res.* United States Environmental Protection Agency, Washington, DC. 1976.
5. Grimsley, J.K.; Scholtz, J.M.; Pace, C.N.; Wild, J.R. *Biochemistry.* 1997, 36, 14366-14374.
6. National Research Council. National Academy of Sciences, Washington, D.C. 1993.
7. Munnecke, D.M.; Hsieh, D.P.H. *Appl. Environ. Microbiol.* 1976, 31, 63-69.
8. Mulbry, W.W.; Kearney, P.C.; Nelson, J.O.; Karns, J.S. *Plasmid.* 1987, 8, 173-177.
9. Dumas, D.P.; Caldwell, S.R.; Wild, J.R.; Raushel, F.M. *J. Biol. Chem.* 1989, 261, 19659-19665.
10. Mulbry, W.W.; Karns, J.S. *J. Bact.* 1988, 171, 6740-6746.
11. Karns, J.S.; Muldoon, M.T.; Derbyshire, M.K.; Kearney, P.C. In Biotechnology in Agricultural Chemistry. LeBaron, H.M.; Mumma, R.O.; Honeycutt, R.C.; Duesing, J.S. ACS Symposium Series no. 334. American Chemical Society, Washington, DC. 1987, 334, 156-170.
12. Lai, K.; Stolowich, N.J.; Wild, J.R. *Arch. Biochem. Biophys.* 1995, 318, 59-64.
13. Dumas, D.P.; Durst, H.D.; Landis, W.G.; Raushel, F.M.; Wild, J.R. *Arch. Biochem. Biophys.* 1990, 227, 155-159.
14. Kolawalski, J.E.; DeFrank, J.J.; Harvey, S.P.; Szafraniec, L.L.; Beaudry, W.T.; Lai, K.H.; Wild, J.R. *Biocat. Biotrans.* 1997, 15, 297-312.

15. Dave, K.I.; Miller, C.E.; Wild, J.R. *Chem. Biol. Interact.* 1993, 87, 55-68.
16. Phillips, J.P.; Xin, J.H.; Kirby, K; Milne, C.P.; Krell, P.; Wild, J.R. *Proc. Natl. Acad. Sci. USA,* 1990, 87, 8155-8159.
17. Steiert, J.G.; Pogell, B.M.; Speedie, M.K.; Laredo, J. *Bio/Technology.* 1989, 7, 65-68.
18. Xu, B.; Wild, J.R.; Kenerley, C.M. *J. Ferm. Bioeng.* 1996, 81, 473-481.
19. Serdar, C.M.; Murdock, D.C.; Rohde, M.F. *Bio/Technology.* 1989, 7, 1151-1155.
20. Serdar, C.M.; Gibson, D.T. *Bio/Technology.* 1985, 3, 567-571.
21. Gill, I.; Ballesteros, A. *Biotechnol. Bioeng..* 2000, 70, 400-410.
22. Caldwell, S.R.; Raushel, F.M. *Appl. Biochem. Biotechnol.* 1991, 1, 59-74.
23. LeJuene, K.E.; Russell, A.J. *Biotechnol. Bioeng.* 1996, 51,450-457.
24. Richins, R.; Kaneva, I.; Mulchandani, A.; Chen, W. *Nature Biotechnol.* 1997, 15, 984-987.
25. Wang, J.; Bhattacharyya, D.; Bachas, L.G. *Biomacromol.* 2001, 3, 700-705.
26. Urry, D.W. *Prog. Biophys. Mol. Biol.* 1992, 57, 23-57.
27. Urry, D.W. *J. Phys. Chem.* 1997, B101, 11007-11028.
28. Shimazu, M.; Mulchandani, A.; Chen, W. *Biotechnol. Bioeng.* 2002, In Press.
29. Rani, N.L.; Lalithakumari, D. *Can. J. Microbiol.* 1994, 40, 1000-1006.
30. Ramanathan, M.P.; Lalithakumari, D. *World J. Microbiol. Biotechnol.* 1996, 12, 107-108.
31. Mulbry, W.W.; Del Valle, P.L.; Karns, J.S. *Pest. Sci.* 1996, 48, 149-155.
32. Mulbry, W.W.; Ahrens, E.; Karns, J.S. *Pest. Sci.* 1998, 52, 268-274.
33. Elashvili, I.; Defrank, J.J.; Culotta, V.C. *Appl. Environ. Microbiol.* 1998, 64, 2601-2608.
34. Hung, S.C.; Liao, J.C. Appl. Biochem. Biotechnol. 1996, 56, 37-47.
35. Shimazu, M.; Mulchandani, A.; Chen, W. *Biotechnol. Prog.* 2001, 17, 76-80.
36. Shimazu, M.; Mulchandani, A.; Chen, W. *Biotechnol. Bioeng.* 2001, 76, 318-324.
37. Wobler, P.K. *Adv. Microb. Physiol.* 1993, 34, 203-235.
38. Mansee, A.; Chen, W.; Mulchandani, A. *Biotechnol. Bioproc. Eng.* 2000, 5, 436-440.
39. Wang, A.A.; Mulchandani, A.; Chen, W. *Appl. Environ. Microbiol.* 2002, 68, 1684-1689.
40. Lai, K.; Dave, K.I.; Wild, J.R. *J. Biol. Chem.* 1994, 269, 16579-16584.
41. Benning, M.M.; Kuo, J.M.; Raushel, F.M.; Holden. H.M. *Biochemistry.* 1995, 34, 7973-7978.

42. Vanhooke, J.L.; Benning, M.M.; Rausel F.M.; Holden, H.M. *Biochemistry.* 1996, 35, 6020-6025.
43. Omburo, G.A.; Kuo, J.M.; Mullins, L.S.; Raushel, F.M. *J. Biol. Chem.* 1992, 267, 13278-13283.
44. diSioudi, B.; Grimsley, J.K.; Lai, K.; Wild, J.R. *Biochemistry.* 1999, 38, 2866-2872.
45. diSioudi, B.; Miller, C.E.; Lai, K.; Grimsley, J.K.; Wild J.R. *Chem. Biol. Interact.* 1999, 119-120, 211-223.
46. Watkins, L.M.; Mahoney, J.J.; McCulloch, J.K.; Raushel, F.M. *J. of Biol. Chem.* 1997, 272, 25596-25601.
47. Sriram, G.; Rastogi, V.; Ashman, W.; Mulbry, W. *Biochem. Biophys. Res. Comm.* 2000, 279, 516-519.
48. Stemmer, W.P.C. *Proc. Natl. Acad. Sci. USA.* 1994, 91, 10747-10751.
49. Stemmer, W.P.C. *Nature.* 1994, 370, 389-391.
50. Zhao, H.; Arnold, F.H. *Curr. Opin. Struct. Biol.* 1997, 7, 480-485.
51. Moore, J.C.; Arnold, F.H. *Nature Biotechnol.* 1996, 14, 458-67.
52. Crameri, A.; Gawes, G.; Rodriguez, E.; Silver, S.; Stemmer, W.P.C. *Nature. Biotechnol.* 1997, 15, 436-438.
53. Bruhlmann, F.; Chen, W. *Abstr. Amer. Instit. Chem. Engineers* 1997 National Meeting.
54. Zhang, J.H.; Dawes, G.; Stemmer, W.P.C. *Proc. Natl. Acad. Sci.* USA 1997, 94: 4504-4509
55. Cho, C.M.; Mulchandani, A.; Chen, W. *Appl. Environ. Microbiol.* 2002, 68, 2026-2030.

Chapter 4

Evolution of New Enzymes and Pathways: Soil Microbes Adapt to *s*-Triazine Herbicides

Lawrence P. Wackett

Department of Biochemistry, Molecular Biology and Biophysics and Biotechnology Institute, 1479 Gortner Avenue, University of Minnesota, St. Paul, MN 55108

Pseudomonas sp. ADP was used as a model to investigate how bacteria evolve new enzymes in new combinations to metabolize *s*-triazine herbicides. The metabolic pathway provides inorganic nitrogen to support microbial growth. The enzymes that initiate the metabolism of the herbicide atrazine are denoted AtzA, AtzB and AtzC. The enzymes are members of the amidohydrolase superfamily and hydrolytically remove substituents from the *s*-triazine ring to generate cyanuric acid. Subsequent to that, cyanuric acid is hydrolyzed by a series of amidases to liberate ring carbon and nitrogen as carbon dioxide and ammonia, respectively. The complete suite of atrazine (*atz*) genes are found on a broad host range plasmid in *Pseudomonas* sp. ADP. The *atz* genes have spread globally and are widespread in different genera of bacteria. These studies begin to reveal how bacterial genes involved in the metabolism of anthropogenic chemicals may arise and spread in the environment.

© 2004 American Chemical Society

Nitrogen heterocyclic compounds are ubiquitous in the biological world (*1*). They are important elements in specialized secondary metabolites and many essential molecules found in all living cells. Microorganisms have been exposed to these compounds for millions of years, require nitrogen to synthesize essential nitrogenous compounds, and thus have developed extensive metabolic mechanisms to catabolize this class. In the soil, these molecules are continually recycled by microbial metabolic activity.

Nitrogenous ring compounds are also important industrially. They are found broadly in chemicals that are designed for environmental applications: antihelmintics, insecticides, herbicides, fungicides and rodenticides (*Table I*). A prominent class of herbicides contain the *s*-triazine ring, a six membered ring with three nitrogen atoms displaced symmetrically in the ring, and two nitrogen-containing substituents on the ring carbon atoms. This class includes atrazine, simazine and cyanazine. To soil microbes, these compounds provide 4-6 nitrogen atoms and are more reactive in nucleophilic aromatic displacement reactions than the biologically familiar pyrimidines (*2*). In this context, it is not surprising that at least some microbes have adapted to use these compounds as the sole source of nitrogen to support growth.

Table I. Commercially-Relevant Nitrogen Heterocyclic Pesticides

Name of Compound	Compound Class	Use
Oxamniquine	Quinoline	Antihelmintic
Imidacloprid	Nicotinoid	Insecticide
Benomyl	Benzimidazole	Fungicide
Nicobifen	Pyridine	Fungicide
Bupirimate	Pyrimidine	Fungicide
Fluoroimide	Pyrrole	Fungicide
Chloroquinox	Quinoxaline	Fungicide
Strychnine	Alkaloid	Rodenticide
Metribuzin	1,2,4-Triazinone	Herbicide
Atrazine	1,3,5-Triazine	Herbicide

s-Triazine herbicides have been remarkably successful because of their ease of manufacture, low toxicity, and long-term effectiveness due to the failure of plants to develop resistance. As a result, herbicides such as atrazine are still important agricultural products more than 40 years after their introduction into the marketplace.

Another issue relevant to herbicide effectiveness, and environmental fate, is the susceptibility of these molecules to microbial breakdown in the environment. The ideal situation is one in which the compound persists long enough to control weeds during some critical period, but is degraded fast enough to not accumulate. If the herbicide is not readily biodegradable, there is the potential for carryover into the next growing season and for human exposure if traces of the herbicide migrate away from the site of application.

s-Triazine Herbicide Metabolism: Early History

The fate of *s*-triazine herbicides, like those of other agricultural chemicals, largely depends on their being metabolized by enzymes in soil microbes. Enzymatic attack may be non-specific or specific. An example of non-specific attack is the oxidative dealkylation of atrazine (*3*), and related compounds, by cytochrome P450 monooxygenases. Oxygenation of alkyl side chain carbons adjacent to a nitrogen atom is the prelude to spontaneous carbon-nitrogen bond cleavage. The organisms which catalyze these reactions are not known to grow on atrazine as a carbon or nitrogen source, and thus this metabolism is thought to be the fortuitous result of non-specific enzymes.

From 1960-1991, a series of papers appeared describing microbial growth on *s*-triazine ring compounds as the sole source of nitrogen (*4-9*). None of those organisms used atrazine and related herbicides, which were considered to be more resistant to metabolism, but rather metabolized simpler structures such as cyanuric acid and melamine. The molecular basis of cyanuric acid metabolism was studied extensively by Cook and coworkers (*Figure 1*) (*10,11*). Enzymatic cyanuric acid hydrolysis was shown to yield biuret. Biuret, in turn, is enzymatically hydrolyzed; urea was detected and proposed to be the product of biuret hydrolase (*Figure 1, top pathway*) (*6*). The enzymatic hydrolysis of urea by the enzyme urease has been known since early in the twentieth century. Urease was purified and crystallized in 1926 (*12*). Microbial metabolism of the industrial compound melamine proceeds through cyanuric acid as an intermediate, and the pathway from that point was thought to be the same as reported for cyanuric acid (*7-9*). The cyanuric acid ring cleavage enzyme (TrzD) was subsequently purified to homogeneity and the reaction product was clearly demonstrated as biuret (*13*). A biuret hydrolase was partially purified by Cook, et al (*6*) but further purification was hindered by the instability of the enzyme activity.

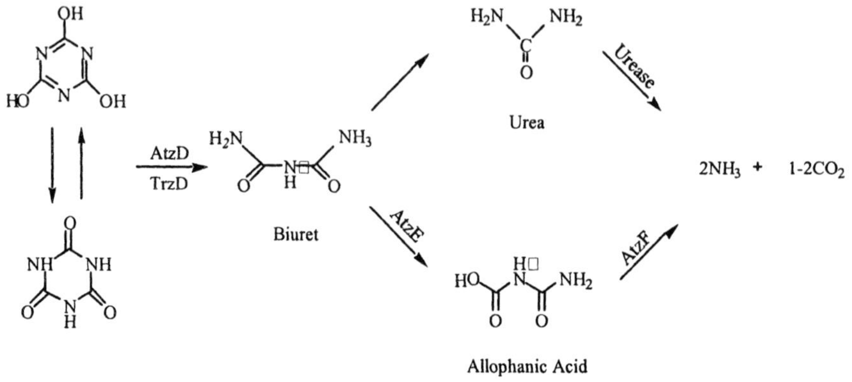

Figure 1. Enzymatic hydrolysis of cyanuric acid via two different pathways.

More recently, the genes involved in cyanuric acid metabolism by *Pseudomonas* sp. ADP were identified, sequenced, and cloned and expressed in *Escherichia coli* (*Figure 1*) (*14*). The cloned cyanuric acid hydrolase produced biuret, consistent with the findings of Karns (*13*). However, the product of another cloned gene in the gene cluster transformed biuret to allophanic acid and the third gene product hydrolyzed allophanic acid to carbon dioxide and ammonia (*Figure 1, bottom pathway*) (*14*). Additional studies with purified enzymes will allow a definitive demonstration of the reactions in the pathway. Moreover, this will reexamine previous results in which the product mixture of the biuret hydrolase reaction was acidified prior to product analysis (*6*). The acid conditions would cause the decarboxylation of allophanic acid, if present, to urea. Thus, one could not distinguish whether urea or allophanate had been formed as an intermediate under the experimental conditions used. *Pseudomonas* sp. ADP, which produces allophanate as an intermediate, will grow on urea, so it is difficult to resolve this point via whole cell studies. The issue will be resolved by purifying the enzymes involved and showing what substrates they process and the products they form under mild conditions in which substrates and products will be reasonably stable.

Atrazine Metabolism to Cyanuric Acid

Cyanuric acid is presumably metabolized by many soil bacteria since it is an intermediate in the metabolism of ammelide which supported growth of a majority of soil bacteria tested (*15*). Thus, metabolic innovations that would lead to the transformation of atrazine to cyanuric acid would permit microbial growth on atrazine as a sole nitrogen source. The first substituent targeted by metabolism, the atrazine chlorine atom, is the most labile to hydrolysis. In fact, soil chemists had detected low concentrations of environmentally formed hydroxyatrazine and proposed that it was derived from soil-catalyzed, abiotic hydrolytic reactions (*16,17*). Hydroxyatrazine did not accumulate in the environment suggesting that it was metabolized by soil microorganisms.

Starting in 1993, different genera of soil bacteria were identified which generate hydroxyatrazine from atrazine (*18-26*). In 1995, de Souza, et al (*27*) cloned a gene region which conferred onto *E. coli* the ability to generate hydroxyatrazine (*Figure 2*). Subsequently, the enzyme atrazine chlorohydrolase was purified to homogeneity (*28*). Gel filtration chromatography indicated an apparent molecular weight of 240,000 for the holoenzyme. The translated protein is predicted to have a molecular weight of 52,421 which, in light of the gel filtration data, suggests a subunit stoichiometry of 4 or 5. The mechanism of hydroxylation catalyzed by atrazine chlorohydrolase was determined using $[^{18}O]$-H_2O. Mass spectrometry confirmed that the product was $[^{18}O]$-hydroxyatrazine and thus the reaction was hydrolytic. More recently, further mechanistic studies have revealed that atrazine chlorohydrolase coordinates ferrous iron, which is essential for its catalytic activity (*29*). Incubating atrazine chlorohydrolase with the chelators 1,10-phenanthroline or oxalic acid removed the required metal with concomitant loss of activity. Chlorohydrolase activity was restored by the addition of Fe(II), Mn(II), or Co(II) salts. Electron paramagnetic resonance (EPR) and electronic spectroscopic studies provided data consistent with a 1:1 metal to subunit stoichiometry. In total, the data indicate that atrazine chlorohydrolase is a metalloenzyme, making this the first report of a metal-dependent hydrolytic dehalogenase.

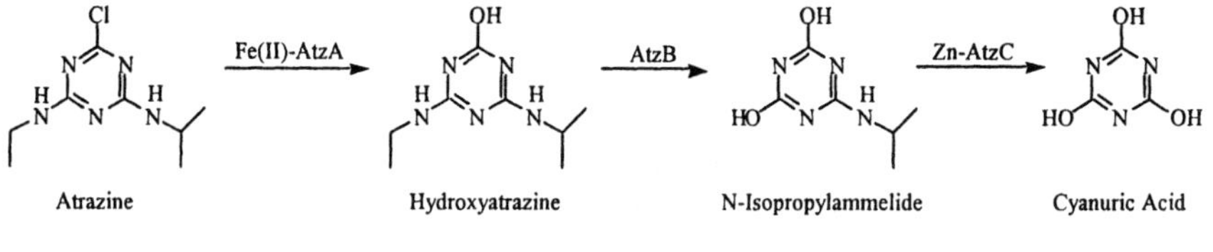

Figure 2. Metabolism of atrazine in Pseudomonas sp. ADP.

Previous studies on other bacterial hydrolytic dehalogenases have revealed different mechanisms and structures. High resolution X-ray structures of haloalkane dehalogenase(*30*), 4-chlorobenzoyl CoA dehalogenase (*31*) and L-2-haloacid dehalogenase (*32*) revealed that these enzymes use an active site aspartate in nucleophilic displacement of the chlorine substituent as chloride anion. The enzyme-substrate ester intermediate is subsequently hydrolyzed by water. These enzymes are evolutionarily related to an alpha/beta hydrolase family of proteins. In contrast, atrazine chlorohydrolase is related to a class of metallohydrolases known as the amidohydrolase superfamily (*33,34*). This protein superfamily includes urease, cytosine deaminase and adenosine deaminase; these enzymes contain essential nickel, iron, and zinc atoms, respectively. Sequence comparisons are consistent with the experimental data indicating the critical role of divalent metal atoms in atrazine chlorohydrolase activity.

Hydroxyatrazine is metabolized by hydroxyatrazine amidohydrolase (AtzB) to yield ethylamine and N-isopropylammelide (*Figure 2*) (*35*). The reaction catalyzed is the hydrolytic removal of a primary alkylamine. More recently, however, a recombinant *E. coli* strain expressing the AtzB enzyme was shown to catalyze a dechlorination reaction (*36*). 2-Chloro-4-hydroxy-6-amino-s-triazine (CAOT) underwent AtzB-dependent dechlorination yielding ammelide as the product. AtzB has yet to be purified so its full substrate range is not rigorously established at this time. Moreover, AtzB is hypothesized to be a metalloenzyme, based on sequence similarity to members of the amidohydrolase superfamily, but this also requires substantiation via purification and characterization of isolated AtzB.

N-isopropylammelide isopropylaminohydrolase, AtzC, the third enzyme in the atrazine degradation pathway in *Pseudomonas* sp. strain ADP, catalyzes the stoichiometric hydrolysis of *N*-isopropylammelide to cyanuric acid and isopropylamine (*Figure 2*) (*34*). More recently, the *atzC* gene has been cloned downstream of the *tac* promoter and overexpressed in *E. coli* to produce AtzC as 36% of the soluble protein (*37*). Purified AtzC hydrolyzed other *N*-substituted aminodihydroxy-s-triazines, and those with linear *N*-alkyl groups showed higher k_{cat} values than those with branched alkyl groups. Like atrazine chlorohydrolase, AtzC required a divalent transition metal for activity. The purified enzyme contained 0.5 equivalents of Zn per subunit. The activity of metal-depleted AtzC was restored by the addition of Zn(II), Fe(II), Mn(II), Co(II), or Ni(II). AtzC is evolutionarily related to members of the amidohydrolase protein superfamily.

AtzA, AtzB and AtzC remove the chloro, N-ethyl and N-isopropyl substituents of atrazine consecutively and specifically (*Figure 2*). For example, AtzC has no detectable activity with hydroxyatrazine. In combination, the three enzymes transform atrazine to cyanuric acid, and thus take a hard-to-degrade compound to a readily metabolizable intermediate.

Insights into the Evolution of a Metabolic Pathway for Atrazine

Acquisition of the *atzA*, *atzB* and *atzC* genes can provide for the ability to use atrazine as a nitrogen source and, potentially, a carbon source. A major question for understanding the evolution of atrazine catabolism is: what is the lineage of the *atzABC* genes? As described above, sequence analysis revealed that the enzymes derive from a protein family that catalyzes metal-dependent hydrolysis reactions. Many of the amidohydrolase superfamily enzymes catalyze the hydrolytic removal of amino groups from heterocyclic ring compounds (*33*). However, the *atz* genes come from distinct origins and have apparently come together in common genomes in recent times (*14,34*). This was immediately apparent from the gross nucleotide composition of the genes. The *atzC* gene has a 39% G+C content while the *atzA* and *atzB* genes are comprised of 58% and 64% G+C, respectively.

The separate origins of the *atzA*, *atzB* and *atzC* genes are also apparent from their genetic localization in *Pseudomonas* sp. ADP. While they are all found on plasmid pADP-1, they are each not directly contiguous to other atrazine genes, but rather are flanked by insertion sequence (IS) elements (*14*). We have obtained deletion plasmids which have spontaneously lost *atzA* or *atzA* plus *atzB* genes, consistent with the idea that the IS elements are active in the movement of the genes. These IS elements share high sequence identity with IS elements flanking other catabolic genes as described by Wyndham, et al. (*38*). Interposed between the *atzB* and *atzC* genes are genes with high sequence identity to those found in the well-characterized mercury resistance operon. The plasmid pADP-1 was shown to confer mercury resistance onto *Pseudomonas* sp. ADP (*14*).

In contrast to the *atzA. atzB*, and *atzC* genes, the *atzD*, *atzE* and *atzF* genes are contiguous on plasmid pADP-1. There is an open reading frame homologous to LysR-type regulatory elements just upstream of the *atzDEF* gene cluster. The LysR-like gene, *atzD*, *atzE*, and *atzF* have similar mol % G+C content (59% to 61%) and codon usage, suggesting that they function as an operon. Consistent with this, there is preliminary evidence that the genes are transcribed as a single mRNA (*14*).

The observations show a clear contrast between the evolutionary history of the *atzABC* genes that take atrazine to cyanuric acid, and the *atzDEF* genes that metabolize cyanuric acid to carbon dioxide and ammonia (*Figure 1*). The lower pathway genes show evidence of residing together for an extensive period. Those genes, or others that are isofunctional, appear to be widespread in the environment. It is fairly common for bacteria to grow on cyanuric acid (*5-8,13*) or ammelide, which is metabolized to cyanuric acid (*15*).

In contrast, the metabolism of atrazine to cyanuric acid is thought to be a very rare phenotype in soil bacteria. The *atzA*, *atzB* and *atzC* genes are not organized into an operonic cluster on plasmid pADP-1. In other atrazine-degrading bacteria, the *atzA*, *atzB* and *atzC* genes are sometimes each found on different plasmids (*39*). A stable consortium that metabolized atrazine was found to have *atzA* and *atzBC* genes in consortium members *Clavibacter michiganese* ATZ1 and *Pseudomonas* sp. CN1, respectively *(40)*. These observations are indicative of a metabolic pathway that is in a recent state of evolutionary flux.

The strongest argument for a recent evolutionary origin of the *atzA, atzB,* and *atzC* genes are the high sequence identity of genes observed amongst atrazine metabolizing bacteria of distinct genera and isolated from four different continents *(41)*. Genes with greater than 99% sequence identity were found in the gram negative bacteria *Pseudomonas, Alcaligenes, Ralstonia, Agrobacterium, Pseudoaminobacter,* and *Chelatobacter* *(18-24)*. More recently, virtually identical *atzB* and *atzC* genes were found in independently isolated gram-positive *Arthrobacter* strains *(25,26)*. These studies are strongly suggestive of the recent evolutionary origin of an upper pathway for atrazine degradation, with a very rapid, broad distribution of the genes into diverse bacteria on a global scale.

The atrazine chlorohydrolase gene appears to encode a unique metabolic activity, the dechlorination of chlorinated s-triazine ring compounds. However, members of the amidohydrolase superfamily have been shown to catalyze fortuitous dechlorination of chlorinated pyrimidines and purines *(43,44)*. In this context, the substrate specificity of atrazine chlorohydrolase (AtzA) from *Pseudomonas* sp. ADP was investigated *(45)*. Purified AtzA catalyzed the hydrolysis of atrazine and fluoroatrazine. What was more significant is that substrate analogs with an amino group or other groups in place of the halogen substituent were not hydrolyzed. Moreover, chlorinated pyrmidines were not substrates. The substrate specificity of AtzA from *Pseudomonas* sp. ADP is very narrow, suggestive of an enzyme that had recently evolved to specifically metabolize *s*-triazine herbicides.

Subsequently, the TriA enzyme, which catalyzes hydrolytic deamination of the aminated *s*-triazine melamine, was found to be 98% identical to AtzA *(46,47)*. While only 9 amino acids out of 475 differed, TriA had insignificant activity with atrazine as a substrate. AtzA shows no activity with aminoatrazine. Thus, TriA and AtzA had completely distinct enzymatic functions despite an amino acid sequence identity of 98%.

Laboratory experiments were conducted to emulate the evolutionary pathway between *triA* and *atzA*, which may have occurred in nature by generating variants that were intermediate in sequence. Site directed mutagenesis was considered to be too cumbersome, over 500 different mutants would need to be made to cover the sequence space between AtzA and TriA. For this reason, DNA shuffling experiments *(48)* were conducted. DNA shuffling produces a combinatorial library of clones, generating proteins having one, two, or three amino acid differences from each parental protein. The clonal library is subsequently screened to discover one or more desirable properties in the new enzyme variants. In this example, a small chemical library was screened to test the breadth of the substrate specificity of each clone. Included in the substrates were atrazine and aminoatrazine, which were distinct and diagnostic substrates for the parental AtzA and TriA enzymes, respectively. The reactivity of each member of the clonal library was screened robotically using high-throughput mass spectrometry (49). The data indicated that the TriA and AtzA enzymes are highly plastic; small amino acid changes had large effects on substrate specificity. Moreover, amino acids at positions 328 and 331 were correlated with the conversion from deaminase to dechlorinase activity. Experiments are currently in progress to test this directly by site directed mutagenesis. The previous results clearly demonstrate that AtzA, an enzyme only known to catalyze dehalogenation, could readily evolve from an enzyme that removes an amino group from a nitrogen heterocyclic ring hydrolytically. The latter activity is quite common in biology. Most organisms contain multiple enzymes catalyzing such reactions; for example, cytosine deaminase and adenosine deaminase. Thus, there were abundant templates of this enzyme class to evolve a dehalogenase that acts on atrazine and related compounds.

Conclusions

The soil chemical milieu is continually changing and microbes must respond to this to survive. A successful response requires the ability to change metabolic capabilities to match new chemical structures that constantly make their way into soil. Enzymes that match up with new structures do not arise *de novo*, but come from pre-existing enzyme libraries encoded by the metagenome of a complex microbial soil flora. The pre-existing enzymes that get recruited for new activities sometimes act on substrates which do not resemble the new substrates, but they draw on common mechanisms to effect a new reaction (50). In the present example, enzymes that metabolize industrial *s*-triazine compounds derive from enzymes that metabolize purine and pyrimidine rings and share the amidohydrolase superfamily fold. Different enzymes from within the amidohydrolase superfamily have been recruited to form a metabolic pathway to efficiently metabolize *s*-triazine herbicides. The genes responsible for these enzymes have been found in different organisms which act together as a metabolic consortium or in single organisms either on different or the same catabolic plasmid. Such systems continually evolve under the changing selective pressures imposed by new chemical inputs into soil.

Acknowledgements

The research described here was the work of many talented students and postdoctorals. This work has been supported by Syngenta Crop Protection, grant

98-35107-6368 from the U.S. Department of Agriculture-NRI/CGP/CSREES, grant 2202-35107-12508 from USDA/CSREES, and grant ER63268-1018305-0007173 from the Office of Science's Office of Biological and Environmental Research, US Department of Energy.

References

1. Stryer, L *Biochemistry*; 4th ed.; W.H. Freeman: New York, NY, 1995.
2. Gilchrist, T. L. *Heterocyclic Chemistry;* Addison Wesley Longman, Harlow: England, 1997; Vol. 17.
3. Nagy, I.; Compernolle, F.; Ghys, K.; Vanderleyden, J.; Demot, R. *Appl. Environ. Microbiol.* **1995**, *61*, 2056-2060.
4. Behki, R.M..; S.U. Kahn. *J. Agric. Food Chem* **1986**, *34*, 746-749.
5. Cook, A.M.; Hütter, R. *J. Agric. Food Chem.* **1981**, *29*, 1135-1143.
6. Cook, A.M.; Beilstein, P.; Grossenbacher, H.; Hütter, R. *Biochem. J.* **1985**, *231*, 25-30.
7. Eaton, R.W.; Karns, J.S. *J. Bacteriol.* **1991**, *173*, 1215-1222.
8. Eaton, R.W.; Karns, J.S. *J. Bacteriol.* **1991**, *173*, 1363-1366.
9. Jutzi, K.; Cook, A.M.; Hütter., R. *Biochem. J.* **1982**, *208*, 679-684.
10. Cook, A.M.; Grossenbacher, H.; Hütter, R. *Biochem. J.* **1984**, *222*, 315-320.
11. Cook, A.M.; Hütter, R. *J. Agric. Food Chem.* **1984**, *32*, 581-585.
12. Sumner, J. B. *J. Biol. Chem.* **1926**, *69*, 435-441
13. Karns, J.S. *Appl. Environ. Microbiol.* **1999**, *65*, 3512-3517.
14. Martinez, B.; Tomkins, J.; Wackett, L.P.; Wing, R.; Sadowsky, M.J. *J. Bacteriol.* **2001**, *183*, 5684-5697.
15. Zeyer, J.; Bodmer, J.; Hütter, R. *Zentralblatt. fur. Bakteriologie,. Mikrobiologie. und. Hygiene.* **1981**, *3*, 289-298.
16. Armstrong, D.E.; Chesters, G. *Environ. Sci. Technol.* **1968**, *2*, 683-689.
17. Li, G.C.; Felbeck, G.T. *Sci.* **1972**, *114*, 201-209.
18. Mandelbaum, R.T.; Wackett, L.P.; Allan, D.L. *App. Environ. Microbiol.* **1993**, *59*, 1695-1701.
19. Mandelbaum, R.T.; Allan, D.L.; Wackett, L.P. *Appl. Environ. Microbiol.* **1995**, *61*, 1451-1457.
20. Radosevich, M.; Traina, S.J.; Yue-Li, H.; Tuovinen, O.H. *Appl. Environ. Microbiol.* **1995**, *61*, 297-302.
21. Topp, E.; Mulbry, W.M.; Zhu, H.; Nour, S.M.; Cuppels, D. *Appl. Environ. Microbiol.* **2000**, *66*, 3134-3141.
22. Topp, E.; Zhu, H.; Nour, S.M.; Houot, S.; Lewis, M.; Cuppels, D. *Appl. Environ. Microbiol.* **2000**, *66*, 2773-2782.
23. Bouquard, C.; Ouazzani, J.; Prome, J.C.; Michel-Briand, Y.; Plesiat, P. *Appl. Environ. Microbiol.* **1997**, *63*, 862-866.

24. Struthers, J.K.; Jayachandran, K.; Moorman,T.B.; *Appl. Environ. Microbiol.* **1998**, *64*, 3368-3375.
25. Rousseaux, S.; Hartmann, A.; Soulas, G. *FEMS Microbiol. Ecol.* **2001**, *36*, 211-222.
26. Rousseaux, S.; Soulas, G.; Hartmann, A. *FEMS Microbiol. Ecol.* **2002**, *41*, 69-75.
27. de Souza, M.L.; Wackett, L.P.; Boundy-Mills, K.L.; Mandelbaum, R.T.; Sadowsky, M.J. *Appl. Environ. Microbiol.* **1995**, *61*, 3373-3378.
28. de Souza, M.L.; Sadowsky, M.J.; Wackett, L.P. *J. Bacteriol.* **1996**, *178*, 4894-4900.
29. Seffernick, J.L.; McTavish, H.; Osborne, J.P.; de Souza, M.L.; Sadowsky, M.J.; Wackett, L.P. *Biochemistry* **2002**, *41*, 14430-14437.
30. Verschueren, K.H.; Franken, S.M.; Rozeboom, H.J.; Kalk, K.H.; Dijkstra, B.W. *J. Mol. Biol.* **1993**, *232*, 856-72.
31. Benning, M. M.; Taylor, K. L.; Liu, R. Q.; Yang, G.; Xiang, H.; Wesenberg, G.; Dunaway-Mariano, D.; Holden, H. M. *Biochemistry* **1996**, *35*, 8103-8109.
32. Ridder, I.S.; Rozeboom, H.J.; Kalk, K.H.; Dijkstra, B.W. *J. Biol. Chem.* **1999**, *274*, 30672-8.
33. Holm, L.; Sander, C. *Proteins.* **1997**, *28*, 72-82.
34. Sadowsky, M.J.; Tong, Z.; de Souza, M.L.; Wackett, L.P. *J. Bacteriol.* **1998**, *180*, 152-158,
35. Boundy-Mills, K.L.; de Souza, M.L.; Wackett, L.P.; Mandelbaum, R.T.; Sadowsky M.J. *Appl. Environ. Microbiol.* **1997**, *63*, 916-923.
36. Seffernick, J.L.; Shapir, N.; Schoeb, M.; Johnson, G.; Sadowsky, M.J.; Wackett, L.P. *Appl. Environ. Microbiol.* **2002**, *68*, 4672-4675.
37. Shapir, N.; Osborne, J.P.; Johnson, G.; Sadowsky, M.J.; Wackett, L.P. *J. Bacteriol.* **2002**, *184*, 5376-5384.
38. Wyndham, R.C.; Cashore, A.E.; Nakatsu, C.H.; Peel, M.C. *Biodegradation* **1994**, *5*, 323-342.
39. Wackett, L.P.; Sadowsky, M.J.; Martinez, B.; Shapir, N. *Appl. Microbiol. Biotechnol.* **2002**, *58*, 39-45.
40. de Souza, M.L.; Newcombe, D.; Alvey, S.; Crowley, D.E.; Hay, A.; Sadowsky, M.J.; Wackett, L.P. *Appl. Environ. Microbiol.* **1998**, *64*,178-184.
41. de Souza, M.L.; Sadowsky, M.J.; Seffernick, J.; Martinez, B.; Wackett, L.P. *J. Bacteriol.* **1998**, *180*, 1951-1954.
42. Strong, L.C.; Pedersen, C.; Johnson, G.; Sadowsky, M.J.; Wackett, L.P. *Appl. Environ. Microbiol.* **2002**, 68, 5973-5980.
43. Baer, H. P.; Drummond, G.I.; Duncan, E. L. *Mol. Pharmacol.* **1966**, *2*, 67-76.
44. Baer, H. P.; Drummond, G.I. *Biochem. Biophys. Res. Comm.* **1966**, *24*, 584-587.

45. Seffernick, J.L.; Johnson, G.; Sadowsky, M.J.; Wackett, L.P. *Appl. Environ. Microbiol.* **2000,** *66,* 4247-4252.
46. Seffernick, J.L.; de Souza, M.L.; Sadowsky, M.J.; Wackett, L.P. *J. Bacteriol.* **2001,** *183,* 2405-2410.
47. Seffernick, J.L.; Wackett, L.P. *Biochemistry* **2001,** *40,* 12747-12753.
48. Stemmer, W.P. *Nature* **1994,** *370,* 389-391.
49. Raillard, S-A.; Krebber, A.; Chen, Y.; Ness, J.E.; Bermudez, E.; Trinidad, R.; Fullem, R.; Davis, C.; Welch, M.; Seffernick, J.; Wackett, L.P.; Stemmer, W.P.C.; Minshull, J. *Chem. Biol.* **2001,** *125,* 1-9.
50. Babbitt, P.C.; Gerlt, J.A. *J. Biol. Chem.* **1997,** *272,* 30591-30594.

Chemically Based Mechanisms

Chapter 5

Detoxification of Some Halogenated Pesticides by Thiosulfate Salts

J. Gan and S. Bondarenko

Environmental Sciences Department, University of California, Riverside, CA 92521

Halogenated organic compounds are used for industrial and agricultural purposes as solvents, degreasing agents, and pesticides. One benign chemical method for remediation and contamination prevention of halogenated compounds in the environment is nucleophilic substitution by thiosulfate salts, in which the compound is dehalogenated and detoxified. This reaction has been shown to occur for methyl bromide (CH_3Br), 1,3-dichloropropene ($C_3H_4Cl_2$), chloropicrin (CCl_2NO_2), methyl iodide (CH_3I), and propargyl bromide (C_3H_3Br), all existing or potential fumigants, and for chloroacetanilide herbicides alachlor, propachlor, acetochlor, and metolachlor. Structural and kinetic analysis suggests that the reaction occurs by S_N2-type nucleophilic substitution, in which thiosulfate ($S_2O_3^{2-}$) replaces the halogen in the molecule. This method has several potential advantages, i.e., rapid dehalogenation rate at ambient temperature, low toxicity of thiosulfate salts, removal of toxicity of parent compounds, and easy availability of common thiosulfate salts. This article is a review of the current state of knowledge about this reaction and its potential applications.

Introduction

Halogenated organic compounds (HOC) have a multitude of uses in modern society. They are widely used in industrial applications as solvent degreasers and refrigerants, in the dry-cleaning industry, in the manufacture of plastics and dyestuffs, and in agricultural applications as active and inactive components of pesticides and fumigants (*1-3*). As a result of the large-scale industrial production and use of HOCs, these compounds have a potential for widespread contamination of natural resources. The Toxic Release Inventory reports releases of more than 1.92×10^8 kg of chlorinated organic compounds in the U.S. for the years 1988 and 1994 to 1996 (*4, 5*).

HOCs are chemicals with a wide range of physical and chemical properties. As a result, the various HOCs differ in their toxicity, persistence, or bioaccumulation potential (*6*). Many of these compounds are toxic or carcinogenic (*3, 7, 8*). Many HOCs resist either chemical or biological attack, which contributes to the environmental persistence of these materials.

A lot of effort is being made to prevent pollution and to restore environmental systems that are already polluted from previous uses (*9*). One of the most attractive methods for remediation of HOCs in ground water and polluted soil is reductive dechlorination by metals such as iron, or zero valent bimetals such as mixtures of zinc-palladium, zinc-nickel, zinc-copper, iron-palladium, iron-nickel, and iron-copper (*10*). Haitko and Eykholt (*11*) reported dechlorination of chlorinated organic compounds by soluble iron citrate. Polychlorinated biphenyls, chlorinated aliphatic hydrocarbons, herbicides and pesticides were successfully dehalogenated by Na/NH_3 or Ca/NH_3 treatment in polluted clay, sandy or organic soils (*12*). The method has several advantages, such as rapid dehalogenation rates at ambient temperature, NH_3 removal from soil, and easy recovery and recycling. Selective mechanochemical dehalogenation of chlorobenzenes was achieved using calcium hydride (*13*). Dechlorination approaches for contaminated soils also include the use of different soil microorganisms (*14, 15*), permanganate oxidation (*16*), new enhanced anaerobic methods (*17*), UV irradiation in the presence of liquid containing oxidation agents (*18, 19*), treatment with polyethylene glycol monomethyl ether potassium salt (*20, 21*), and reaction with sulfite (*22*). Many of these treatments, however, present some problems for decontaminating polluted natural media, and the disadvantages include limited breadth in reactivity, expensive reagents, toxic reaction products, and nonselective reactions. There is thus a need to develop new and selective approaches to decontaminate HOCs in the environment.

A HOC containing sp^3-hybridized carbon-halogenated bonds can undergo reaction by several routes: nucleophilic substitution, dehydrohalogenation (loss of HX; X = halogen), and reductive dehalogenation. Reductive dehalogenation can react via either hydrogenolysis (replacement of halogen by hydrogen) or reductive elimination (loss of two vicinal or geminal halogens). It is known that mono- and dihalogenated organic compounds, which contain sp^3-hybridized carbon-halogenated bonds, are capable of bimolecular nucleophilic substitution (S_N2) reactions *(23, 24)*. For more highly substituted HOCs, dehydrohalogenation and reductive dehydrohalogenation predominate over S_N2 reactions.

Taking into account these facts, our group has been exploring the use of thiosulfate salts to dehalogenate HOCs since 1997. So far, we have focused our research effort on halogenated pesticides. The reaction is S_N2-type nucleophilic substitution, in which thiosulfate ($S_2O_3^{2-}$) replaces halogen in the HOC molecule *(9, 25-29)*. This reaction has been shown to occur with methyl bromide (CH_3Br), 1,3-dichloropropene ($C_3H_4Cl_2$), chloropicrin (CCl_2NO_2), methyl iodide (CH_3I), and propargyl bromide (C_3H_3Br), all existing or alternative soil fumigants, and with chloroacetanilide herbicides including alachlor (2-chloro-2',6'-diethyl-*N*-methoxy-methyl-acetanilide), propachlor (2-chloro-*N*-isopropylacetanilide), acetochlor (2-chloro-*N*-ethoxymethyl-6'-ethylacate-o-toluidide), and metolachlor [(2-chloro-6'-ethyl-*N*-(2-methohy-1-methylethyl) acet-o-toluidide)]. These chemicals have extremely heavy use worldwide. For instance, the combined use of these chemicals in the United States alone exceeds 1.0×10^8 kg each year. The general form of the reaction is as given below,

$$RX + S_2O_3^{2-} \longrightarrow RS_2O_3^- + X^-,$$

where X can be Cl, Br, or I.

The use of thiosulfate salts has several advantages. First, thiosulfate is one of the most powerful nucliophiles *(30)*. Secondly, thiosulfate salts are low in toxicity and relatively safe to use. For example, at 2.5 g/kg, the LD_{50} for rats of sodium thiosulfate is similar to that of sodium chloride (3.75g/kg). Lastly, thiosulfates are available at low cost, because, common thiosulfate salts, such as ammonium thiosulfate (($NH_4)_2S_2O_3$), potassium thiosulfate ($K_2S_2O_3$), and calcium thiosulfate (CaS_2O_3), are commercial fertilizers or soil amendments. The other thiosulfate salt, sodium thiosulfate ($Na_2S_2O_3$), is also easily available, as it is heavily used in the photographic industry.

This review is a summary of our research findings over the last few years in characterization and application of this reaction.

Transformation Kinetics in Solution

Transformation of HOCs by thiosulfate was first studied in the aqueous phase. It was shown that as the molar ratio of HOCs to thiosulfate salt in water increased, the dissipation of the HOCs proportionally accelerated *(25, 26)*. For example, in solutions containing ammonium thiosulfate at 2.0 m*M*, the first-order half-life ($T_{1/2}$) of methyl iodide decreased from 7250 h (in water) to 8.5 h and that propargyl bromide decreased from 3100 h (in water) to 9.8 h. When the ammonium thiosulfate concentration was further increased to 8 mM, the $T_{1/2}$ of methyl iodide and propargyl bromide was only 1.4 and 1.6 h, respectively (Figure 1). The effect of different thiosulfate concentrations on the disappearance of a chloroacetanilide herbicide is illustrated with alachlor in Figure 2.

The second-order reaction constant k was calculated for fumigants by fitting the measured data to the second-order kinetic model (Table 1). Good correlation was found for all fumigants except for chloropicrin, with correlation coefficients r ≥0.99 (Table 1). This suggests that the reaction between HOCs and thiosulfate salt in aqueous solution follows second-order kinetics typically exhibited by S_N2 reactions.

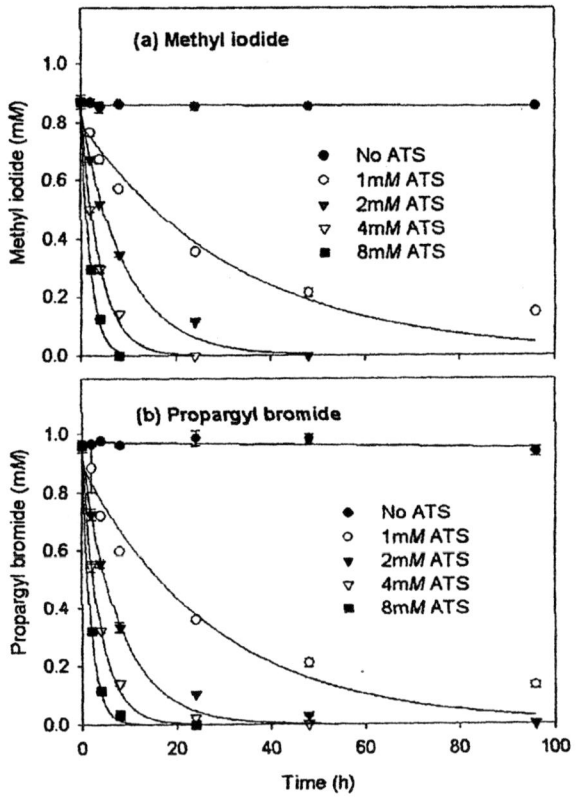

Figure 1. Dissipation of fumigants (1mM) in thiosulfate solutions: (a) methyl iodide and (b) propargyl bromide (28)

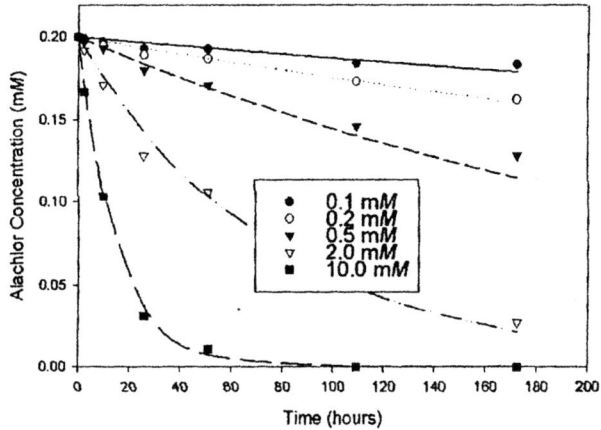

Figure 2. Dissipation of alachlor (0.2 mM) in aqueous solutions with different initial concentrations of ammonium thiosulfate (29)

Table 1. Reaction rate constant k ($M^{-1}s^{-1}$) and regression coefficient r of fumigants and ammonium thiosulfate in aqueous phase at 20°C (28)

Fumigant	k	R
Methyl bromide	2.11×10^{-2}	1.00
Methyl iodide	1.87×10^{-2}	1.00
Propargyl bromide	1.59×10^{-2}	0.99
Chloropicrin	3.5×10^{-4}	0.70
1,3-dichloropropene	1.95×10^{-3}	0.99

Reaction Pathways

Analysis of reaction mixtures by ion chromatography (IC) showed that as each HOC reacted with thiosulfate, X^- ($X^- = Cl^-$, Br^-, or I^-) was liberated into the solution. The rate of X^- release always equaled the rate of HOC consumption, as shown in Figure 3 for propargyl bromide and Figure 4 for alachlor. This analysis provided evidence that X^- was displaced from the HOC molecule during the reaction, indicating that the reaction was a stoichiometric substitution. However, in the transformation of chloropicrin, consumption of $S_2O_3^-$ was four times greater than that of chloropicrin, while accumulation of Cl^- was about two times faster than the dissipation of chloropicrin (*28*). It is likely that more than one of three Cl atoms were replaced by $S_2O_3^-$ in the transformation or that other reactions occurred that consumed additional $S_2O_3^-$.

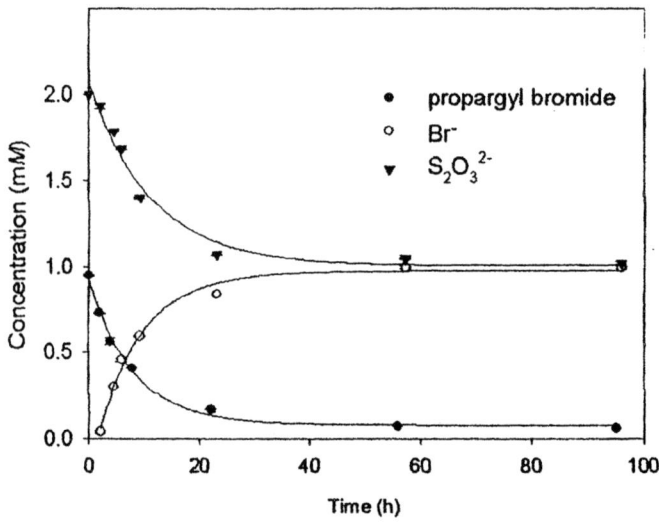

Figure 3. Consumption of propargyl bromide (1mM) and thiosulfate (2mM) and accumulation of bromide in aqueous phase during propargyl bromide transformation by thiosulfate salt (28)

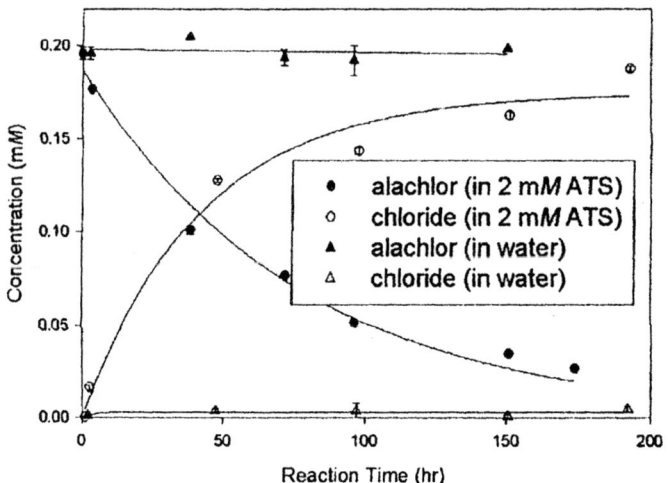

Figure 4. Consumption of alachlor (1mM) and thiosulfate (2mM) and accumulation of chloride in aqueous phase during alachlor transformation by thiosulfate salt (29)

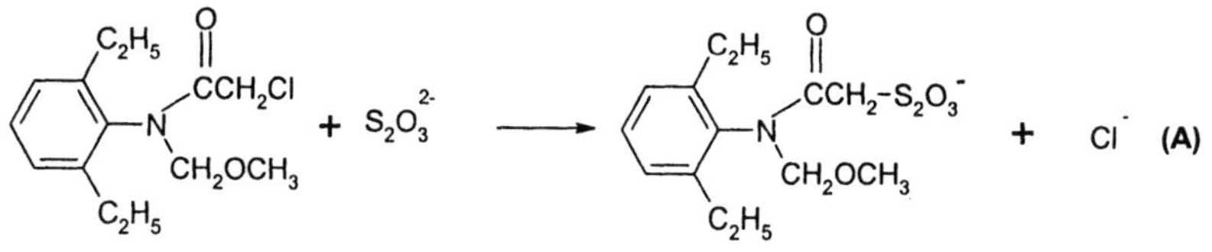

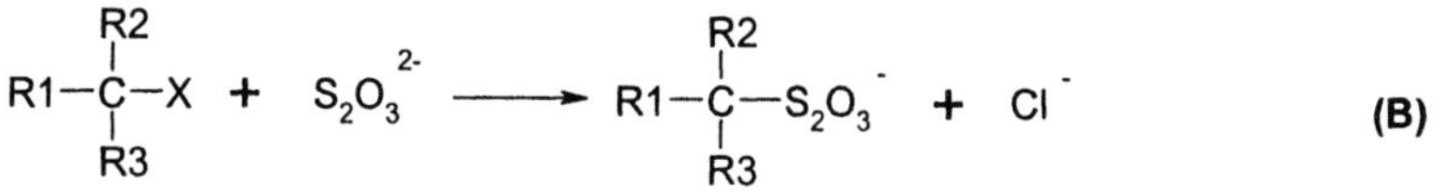

Figure 5. The initial step of nucleophilic dehalogenation reaction between alachlor and thiosulfate salts (A) and fumigants and thiosulfate salts (B)

Since the reaction between HOCs and thiosulfate followed second-order kinetics, and that X^- and dehalogenated HOC-thiosulfate derivative was formed as one of the initial products, it may be concluded that the reaction followed a pathway as depicted for alachlor in Figure 5A or for fumigants in general in Figure 5B.

Relative Reactivity

The relative reactivity was compared under the same conditions for fumigants, where the initial concentration of fumigants was 1 mM and the initial concentration of ammonium thiosulfate was varied from 0 to 10 mM. The rate of fumigant dissipation followed an order of methyl bromide ≈ methyl iodide > propargyl bromide > 1.3-D ≈ chloropicrin. The relative reactivity among the fumigants is shown also in the difference in the second-order rate constants (Table 1). The relative reactivity of these compounds with thiosulfate may be explained by the steric hindrance of substitution groups on the primary carbon and the tendency for the leaving group to leave during nucleophilic substitution. As the nuchleophile would attack a HOC molecule from the direction opposite to the leaving group, a bulky substitution on the primary carbon would prevent easy approaching of the incoming nucleophile, rendering the reaction slower. Thus, methyl bromide and methyl iodide are more susceptible to nucleophilic attack. In addition, Br and I are better leaving groups than Cl. The reactivity of the herbicides decreased in an order of propachlor > alachlor > acetochlor > metolachlor (Figure 6). This dependence suggests again that the substitutions at the nucleophilic center, i.e., the chlorinated carbon, influenced the reactivity. The bulky substitutions at this position may have resulted in greater steric hindrance for metolachlor, rendering its primary carbon less accessible by $S_2O_3^{2-}$.

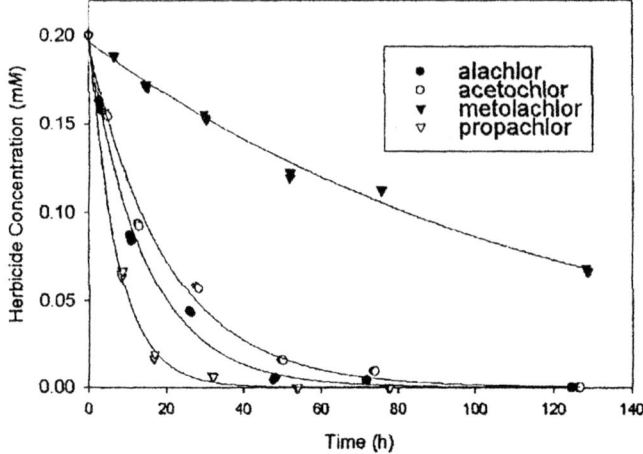

Figure 6. Disappearance of chloroacetanilide herbicides in 10 mM thiosulfate solutions at 20° C (29)

Enhanced Transformation in Soil

The reaction was further tested in soils when a known rate of ATS was added to soil with selected fumigants. Greatly enhanced transformation of fumigants was consistently observed. For instance, first-order $T_{1/2}$ for methyl bromide transformation was 1300 h in a unamended sandy loam soil, but was reduced to 21 h when ATS was added to the soil at 1.0 mmol kg^{-1} (Table 2). Under the same conditions, the persistence of the other fumigants decreased by factors of 5-65. The reduction was greater for methyl bromide, methyl iodide, and propargyl bromide as compared to 1,3-D or chloropicrin (Table 2). Comparing fumigant dissipation in sterile and nonsterile soils showed that the enhanced degradation in thiosulfate-amended soils was a chemically derived transformation (*25, 26*). Comparing thiosulfate-induced fumigant transformation in different types of soil showed that the enhancement was similar in different soils. Thiosulfate-induced fumigant transformations were consistently accelerated as the temperature increased. This temperature dependence implies that the reaction would be more effective when the soil is warm, a condition that may be created under plastic covers that are commonly used in soil fumigation (*25, 26*). Thiosulfate-induced fumigant transformation in soil was relatively insensitive to moisture variations within the normal soil moisture range (*27*). These findings imply that thiosulfate salts may be introduced into soil to accelerate fumigant transformation, and the application should be effective under common field conditions.

Table 2. Half-lives (h) of fumigants (0.5 mmol kg^{-1}) in Arlington Sandy Loam with and without addition of ammonium thiosulfate (ATS) at 1.0 mmol kg^{-1} (28)

Fumigant	ATS-amended		Blank control	
	$T_{1/2}$	R	$T_{1/2}$	r
Methyl bromide	20.5	0.99	1300	0.53
Methyl iodide	23.5	0.98	495	0.91
Propargyl bromide	12.9	0.97	290	0.93
Chloropicrin	30.1	0.97	139	0.95
1,3-dichloropropane	29.7	0.98	162	0.99

The reaction was also tested in soils and sand for chloroacetanilide herbicides. Enhanced herbicide dissipation occurred in sand or sandy soils, but became insignificant in clayey soils (29). The reduced enhancement in clayey soils may be attributed to the strong adsorption of herbicides to soil, which may have limited the interaction between the herbicide (adsorbed) and thiosulfate anion (in solution). In sand or sandy soils, the transformation of propachlor was the most rapid, which was followed by acetochlor and alachlor. Metolachlor transformation was only marginally enhanced.

Detoxification

Bacteria-based bioassays were used to evaluate changes in toxicity of the HOCs caused by the reaction. The acute toxicity test was carried out using a illumescent bacterium *Vibrio fisheri*. The measurement is based on the principle that the emitted illuminescence from the bacteria is suppressed under toxicity stress, and EC$_{50}$ can be obtained from the relationship of illuminescene and the concentration of test compound. Bacterial EC$_{50}$ increased significantly after the fumigants were reacted with sodium or ammonium thiosulfate, suggesting that the acute toxicity greatly decreased after the reaction (Table 3). For instance, bacterial toxicity decreased by nearly 200 times for chloropicrin, >300 times for methyl iodide, and >700 times for 1,3-dichloropropene (Table 4). The bacterial EC$_{50}$ increased from 300-550 mM for herbicides to 1,360-4,400 mM after transformation by thiosulfate salts. When the bacterial EC$_{50}$ is >1,500 mM, the test compound may be considered nontoxic or only slightly toxic. Therefore, it may be concluded that transformation of herbicides by thiosulfate is also a detoxification process.

The biological activity of many HOCs correlates with their alkylating ability. The activity arises from direct alkylation of critical biological molecules by fumigants. It is likely that transformation by thiosulfate removes the nucleophilic center (i.e., Cl, Br, or I) from the HOC, thus preventing the HOC from reacting further with biological systems. These preliminary assays suggest that thiosulfate-induced dehalogenation is a detoxification reaction, and the use of this reaction would be environmentally compatible. This, when combined with the easy availability of common thiosulfate salts, implies that the use of this reaction for environmental decontamination holds great promises.

Table 3. Changes in bacterial EC_{50} values of fumigants before and after reaction with ammonium thiosulfate in aquoues phase (28)

Fumigant	$EC50$ (mM) Fumigant solution	$EC50$ (mM) Reacted solution	Difference (times)
Methyl bromide	1.22	4.38	3.6
Methyl iodide	2.95	>1000	>300
Propargyl bromide	0.38	2.58	6.7
1,3-dichloropropene	0.13	>100	>700
Chloropicrin	9.4×10^{-4}	0.18	191
Ammonium thiosulfate	248		

Table 4. Changes in bacterial EC_{50} values of herbicides before and after reaction with thiosulfate salt in aquoues phase

Herbicide	$EC50$ (mM) Herbicide solution	$EC50$ (mM) Reacted solution	Difference (times)
Alachlor	494	4376	8.9
Propachlor	358	1361	3.8
Acetochlor	297	1839	6.2
Metholachlor	544	1395	2.6
Phenol	5177		

Application Examples

Fumigant Emissions Reduction

Methyl bromide is known for its very high volatility. Methyl bromide emission during soil fumigation has been suggested to contribute to stratospheric ozone depletion. For this reason, methyl bromide is undergoing phase out in the United States and some other countries. In our earlier studies, we explored the use of surface amendment of thiosulfate fertilizers to reduce methyl bromide emission by neutralizing it before it is emitted into the atmosphere. It was observed that rapid CH_3Br volatilization occurred shortly after it was injected into the soil columns (25). However, in thiosulfate-amended soil, the magnitude of volatilization fluxes was greatly reduced as compared to the control column. For example, the maximum flux detected from the control column without thiosulfate amendment was 1,400 µg h^{-1}, while that from the thiosulfate-amendment column was only about 300 µg h^{-1}. The overall emission loss of methyl bromide was 61% from the control treatment, but decreased to only about 8% in the ammonium thiosulfate amended columns (25). In preliminary field tests, no adverse effect on pesticide control efficacy was observed when ammonium thiosulfate was applied to the soil surface. It is likely that methyl bromide transformation was limited only to the surface soil layer, and the depletion of methyl bromide did not compromise the control of pesticides dwelling in the root zone below the soil surface.

Surface amendment of thiosulfate products was also tested for reducing 1,3-D emission after soil treatment using packed soil columns or field plots (27). In packed soil columns, it was found that the reduction in 1,3-D emission was proportional to the amount of ammonium thiosulfate used when the amount of water (as carrier for ammonium thiosulfate) was fixed, and to the amount of water sprayed when the amount of ammonium thiosulfate was fixed (Table 5). When ammonium thiosulfate was used at 64 g m^{-2}, the emission of 1,3-D decreased from 33% to 12% when the amount of water was increased from 1 mm to 9 mm (Table 5). When the water application rate was kept at 9 mm, 1,3-D emission loss decreased from 15% to only 3% when the rate of ammonium thiosulfate was increased from 64 g m^{-2} to 193 g m^{-2}. Under similar conditions, the total emission loss of 1,3-D from untreated columns was 43-47%. In field plots, the emission loss of 1,3-D was observed to decrease from 25% in unamended plots to about 5% in ammonium thiosulfate-amended plots. In addition, application of either ammonium or potassium thiosulfate to field plots did not significantly inhibit the effectiveness for nematode control.

Soil Remediation

In subsequent studies, we also tested the use of sodium or ammonium thiosulfate to remove HOCs from soil or sand, with the objective for soil or aquifer remediation. Small soil columns were packed with sandy soil or sand, and contaminated with herbicides at a known concentration. In one group of columns, ammonium thiosulfate was introduced into the column, and in another group of column, only water was injected into the column. The columns were then allowed to sit at ambient temperature for about a week. Water (0.01 M

Table 5. Total emissions (% of applied) of 1,3-D from soil columns following surface application of ammonium thiosulfate (ATS) in different amounts of water and at different rates

Treatment	cis (%)	trans (%)	Mean (%)	Reduction (%)[†]
Water application rate:				
1,3-D only	43.1	35.1	38.2	-
1,3-D + ATS in 1mm water[‡]	33.4	30.1	31.0	- 19%
1,3-D + ATS in 3mm water	23.1	26.0	23.9	- 37%
1,3-D + ATS in 9mm water	12.4	16.7	14.2	- 63%
ATS application rate:				
1,3-D only	47.5	40.2	42.9	-
1,3-D + 1mL Thio-Sul[§]	14.5	19.6	16.6	- 61%
1,3-D + 2mL Thio-Sul	6.4	10.9	8.4	- 80%
1,3-D + 3mL Thio-Sul	3.4	6.6	4.9	- 89%

[†] Calculated as reduction from the 1,3-D only treatment.

[‡] 1 mm in depth = 11 ml in volume.

[§] 1 ml Thio-Sul = 64 g m^{-2} as ammonium thiosulfate (ATS).

Source: Reproduced with permission from reference 27. Copyright 2000 John Wiley and Sons

$CaCl_2$) was then pumped into the columns to leach the herbicides, and the leachate was collected in fractions and quantified for herbicide concentration on HPLC. Addition of ammonium thiosulfate to the saturated soil or sand columns essentially prevented propachlor or alachlor from appearing in the leachate, achieving ~100% removal of the herbicide as compared with the untreated columns (Figure 7). This simple demonstration indicates that thiosulfate salts may be introduced into polluted soil or aquifer systems to detoxify HOC contaminants that are otherwise difficult to remove. The introduction of thiosulfate salts may be easily achieved by injection through wells or by leaching in through controlled surface irrigation.

References

1. Golfinopoulos, S.K.; Lekkas, T.D.; Nilolaou, A.D. *Chemosphere.* **2001**, *45*, 275-284.
2. Aga, D.; Thurman, E.M. *Environ. Sci. Technol.* **2001**, *35*, 2455-2460.
3. Stackelberg, P.E.; Hopple, J.A.; Kauffman, L.J. 1997. U.S. Geological Survey Water-Resources Investigations Report 97-4241, USGS, Denver, CO.

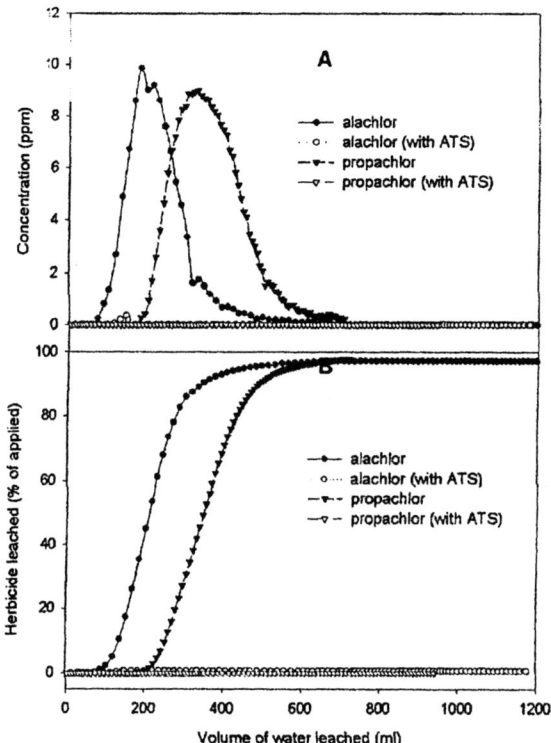

Figure 7. Effect of ammonium thiosulfate- amendment on alachlor leaching through a sand column and propachlor leaching through a soil column. (A) Herbicide concentrations in leachate. (B) Cumulative fractions of herbicides leached through (29)

4. U.S. Environmental Protection Agency. 1996a. Toxics Release Inventory Public Data Release. EPA 745-R-98-005, EPA, Washington, D.C.
5. U.S. Environmental Protection Agency. 1996b. Toxics Release Inventory State Fact Sheets. EPA 745-F-98-00a, EPA, Washington, D.C.
6. Lee, M. D.; Davis, J.W. In *Natural Remediation of Environmental Contaminants: Its Role in Ecological Risk Assessment and Risk Management;* Swindoll, M.; Stahl R.G.; Ells, S.J., Eds.; SETAC Press: Pensacola, FLA, 2000; pp 199-245.
7. Masters, G.M. 1998. *Introduction to Environmental Engineering and Science.* Prentice-Hall: Upper Saddle River, NJ.
8. Sacks, R.; Akard, M. *Environ. Sci. Technol.* **1994**, *28*, 428A-433.
9. Gan, J.; Yates, S.R.; Becker, J.O.; Knuteson, J. *J. Environ. Qual.* **2000**, *29*, 1476-1481.

10. Kim, Y.-H.; Carraway, E.R. *217th ACS National Meeting Abstracts*. ACS: Washington, D.C., 1999; pp 137.
11. Haitko, D.A.; Eykholt, G.R. U.S. Patent 5,575,926, 1996.
12. Pittman, C.U.; He, J.J. *J. Haz. Mat.*. **2002**, *92*, 51-62.
13. Loiselle, S.; Branco, M.; Mulas, G.; Cocco, G. *Environ. Sci. Technol.* **1997**, *31*, 261-265.
14. Klag, P.; Weltzin, M.; Eisentrager, A.; Dott, W.; Ruden, H. *Acta Biotechnologica.* **2001**, *21*, 129-139.
15. Doong, R.-A.; Wu, S.-C. *Water Res.* **1996**, *30*, 577-586.
16. Poulson, S.R.; Naraoka H. *Environ. Sci. Technol.* **2002**, *36*, 3270-3274.
17. Feldman, S.M.; Lobasso, T.; Burdick, J.S.; Standish, R. In *Proceedings from WASTECON 2000, SWANA's Annual International Solid Waste Exposition;* Miller A.G., Ed.; Solid Waste Association of North America: Silver Spring, OH, 1999; pp 135-148.
18. Haginoya, S.; Arai, T. Japan Patent 2,000,005,562, 2000.
19. Kitchens, J.A. U.S. Patent 4,144,152, 1974.
20. King, A.B.; Hoch, R. *Proc. Annu. Meet. – Air Waste Manage. Assoc.* **1991**, *11*, 9-14.
21. Paleologou, M.; Purdy, W.C.; Misra, S.K.; Korczak, S.Z. *Int. J. Environ. Anal. Chem.* **1993**, *50*, 215-242.
22. Bauman, L.; Stenstrom, M.K. *Environ. Sci. Technol.* **1989**, *23*, 232-236.
23. Lippa, K.A.; Roberts, A.L. *Environ. Sci. Technol.* **2002**, *36*, 2008-2018.
24. Terney, A.L. *Contemporary Organic Chemistry;* W.B. Saunders Company: Philadelphia, London, Toronto, 1979.
25. Gan, J.; Yates, S.R.; Becker, J.O.; Wang, D. *Environ. Sci. Technol.* **1998**, *32*, 2438-2441.
26. Gan, J.; Papiernik, S.R.; Yates, S.R.; Jury, W.A. *J. Environ. Qual.* **1999**, *28*, 1436-1441.
27. Gan, J.; Becker, J.O.; Ernst, F.F.; Hutchinson, C.; Knuteson, J.A.; Yates, S.R. *Pest. Manag. Sci.* **2000**, *56*, 264-270.
28. Wang, Q.; Gan, J.; Papiernik, S.R.; Yates, S.R. *Environ. Sci. Technol.* **2000**, *34*, 3717-3721.
29. Gan, J.; Wang, Q.; Yates, S.R.; Koskinen, W.C.; Jury, W.A. *PNAS (USA).* **2002**, *99*, 5189-5194.
30. Schwarzenbach, R.P.; Gschwend, P.M.; Imboden, D.M. *Environmental Organic Chemistry;* Jonh Wiley and Sons.: Chichester-Brisbane, Toronto, Singapore, 1991.

Chapter 6

Anodic Fenton Degradation of Pesticides

A. T. Lemley[1], Q. Wang[1], and D. A. Saltmiras[2]

[1]Graduate Field of Environmental Toxicology, Cornell University, Ithaca, NY 14853-4401
[2]Ashland Inc., 5200 Blazer Parkway, Dublin, OH 43017

An electrochemical adaptation of the Fenton treatment method has been successfully applied to the degradation of pesticides and other contaminants in aqueous solution. The results of this work have implications for fast and efficient treatment and pretreatment of aqueous wastes. The anodic Fenton treatment (AFT) system has been applied to the treatment of ethylene thiourea, trifluralin, atrazine, 2,4-D, diazinon, carbaryl, carbofuran, and other carbamate pesticides. The method was initially developed in a batch system that can be scaled up to a flow-through system as a convenient way to deliver Fenton reagents. Ferrous ion is produced at an iron anode while hydrogen peroxide is delivered via a peristaltic pump. Separation of the anode chamber from the inert cathode (where water is reduced), first by a salt bridge and then by an anion exchange membrane, improved the method considerably, making it as effective as traditional Fenton chemistry without the disadvantages. Development of a kinetic delivery model for the AFT method has made it an extremely useful probe to study hydroxyl radical reactions rates using competitive kinetics. These advances, coupled with the use of GC-MS to identify degradation products, has enabled the study of reaction mechanisms for degradation of these aqueous contaminants.

© 2004 American Chemical Society

Introduction

Evidence that water soluble pesticides are significant groundwater and surface water contaminants continues to emerge in nationwide and geographically specific studies. The breadth of pesticide contamination of groundwater is shown in the comprehensive study by the US Geological Survey (1). Recent reviews have continued to emphasize that the handling and disposal of wastewater is a significant pesticide waste management issue in the US (2,3). These same pesticides and their degradation products are potential contaminants with normal agricultural use. Typically, hundreds of gallons of water with low concentrations of active ingredient are generated at an agricultural facility. Simple calculations show that these wastes represent many pounds of active ingredient which are left over each season and which must be stored and/or eventually treated.

As a major advanced oxidation technology, the Fenton reaction (eq. 1) has received extensive attention in the past decade for degradation of environmental contaminants, including pesticides (4-14).

$$Fe^{2+} + H_2O_2 = Fe^{3+} + OH^- + {^\cdot}OH \tag{1}$$

The hydroxyl radical (·OH) generated from the Fenton reaction is found to be a strong oxidant to many toxic or non-biodegradable organics in wastewater. The classic Fenton reaction and an electrochemical Fenton approach have been shown to be effective with respect to the degradation of organophosphorus pesticides: methyl parathion, malathion, and methamidophos (15,16); triazines: atrazine and cyanazine; chloroacetamides: alachlor and metolachlor; and the substituted picolinic acid, picloram (11). This method has the advantage of controlling the reaction by reactant delivery and avoiding the handling of large amounts of ferrous salt. It also provides effluent at circum-neutral pH, an environmental positive, but a negative with respect to efficiency. To overcome this difficulty, anodic Fenton treatment (AFT) was proposed as an improvement to electrochemical Fenton treatment and was used to study the degradation of ethylene thiourea (ETU), trifluralin and atrazine (14, 17, 18). This paper will review the successful application of AFT to a variety of pesticides at concentrations found in wastewater, continued development and improvement of the apparatus, the development of a kinetic delivery model, study of competitive kinetics, and preliminary evaluation of AFT as a pretreatment for microbial degradation methods. Work in the future will evaluate the feasibility of applying this technology to pesticide contamination at lower concentrations in groundwater.

Materials and Methods

Chemicals

Pesticides (usually 99%) were purchased from Chem Services (West Chester, PA 19381). Chemical structures of pesticides for which data are reported and which are not shown in other figures are shown in Figure 1. Hydrogen peroxide (analytical grade), acetonitrile (HPLC grade), and water (HPLC grade) were purchased from Mallinckrodt (Paris, KY 40361). Sodium chloride (certified), hydrate ferrous sulfate (certified), phosphoric acid (analytical grade), and hexanes (HPLC grade) were purchased from Fisher Scientific (Fair Lawn, NY 07410). The anion exchange membrane (ESC-7001) and cation exchange membrane (ESC-7000) used in some experiments were purchased from Electrosynthesis (Lancaster, NY14086). Properties of the membranes are available, but composition is not.

Treatment System

The treatment system consists of two half-cells, which are connected via a salt-bridge (containing saturated NaCl) or an anion exchange membrane. The ferrous ion is delivered into the anodic half-cell by electrolysis from an iron anode (eq. 2). Hydrogen peroxide is constantly added to the anodic half-cell by a peristaltic pump. In the cathodic half-cell, water is reduced on a graphite cathode (eq. 3). The most significant advantages of AFT over other Fenton treatment technologies are that the Fenton reaction occurs in self-developed optimal acidic conditions (pH~3), and the pH of the treatment effluent can be partially neutralized by combining the solutions from the two half-cells. At the same time it retains the advantages of the electrochemical system with respect to controlled delivery and easy handling of reactants.

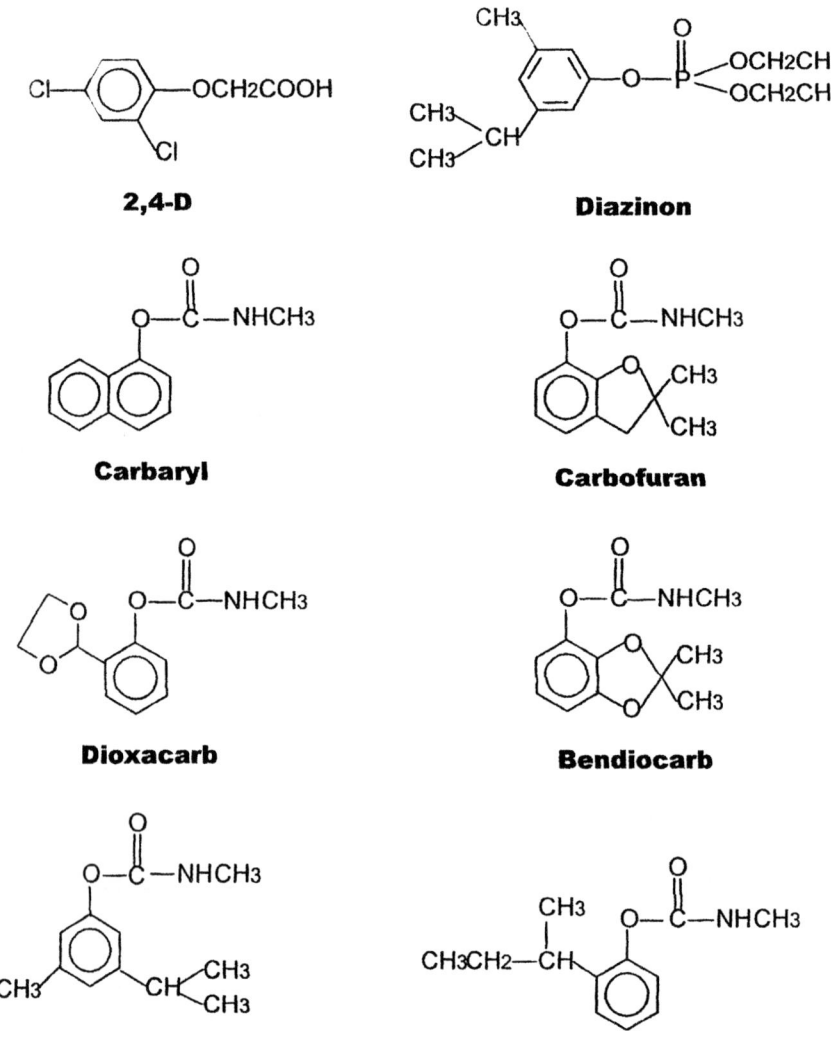

Figure 1. Chemical structures of 2,4-D, diazinon, carbaryl, carbofuran, dioxacarb, bendiocarb, promecarb, and fenobucarb.

Anode: $Fe \rightarrow Fe^{2+} + 2e^-$ (2)

Cathode: $2H_2O + 2e^- \rightarrow H_2 + 2OH^-$ (3)

Degradation experiments were carried out in two 300-mL glass half-cells that served as anodic and cathodic half-cells. These two half-cells were separated by either a salt bridge or an anion exchange membrane. Typically 200 mL of 100 µM pesticide solution with 0.02 M NaCl was added into the anodic half-cell, and the same volume of 0.08 M NaCl aqueous solution was added into the cathodic half-cell. Each of the two half-cells was stirred using a magnetic stirring bar. A 2 cm× 10 cm× 0.2 cm iron plate and a 1 cm (i.d.)× 10 cm(L) graphite rod were used as anode and cathode, respectively. The electrolysis current was supplied by a BK Precision® DC power supply 1610 and was controlled at 0.050 Amp. Hydrogen peroxide solution of 0.3109 M (prepared by diluting from a solution of known concentration) was delivered into the anodic half-cell using a Fisher Scientific peristaltic pump at a rate of 0.50 mL·min^{-1}. The delivery rate of ferrous ion was calculated from the electrolysis current than can be adjusted through the power supply. The typical ratio of H_2O_2: Fe^{2+} was 10:1. The temperature was kept at 24.7±0.5 °C. The electrolysis was started by turning on the power supply when the first drop of hydrogen peroxide entered the anodic half-cell. At different time intervals 1.00 mL of anodic solution was removed from the anodic half-cell and added into a 2-mL GC-vial containing 0.10 mL methanol (to quench the subsequently generated hydroxyl radical) for HPLC analysis. Treatments were repeated two times for a total of three replications.

Analytical Methods

In general, pesticides were analyzed using an HP 1090 HPLC equipped with a Diode Array Detector. For atrazine, the column was a Restek Octylamine, 150 x 4.6mm with 3µm particle size with 1cm guard cartridge and 2µm frit, and the mobile phase consisted of solvent A (water with pH adjusted to 3 with HCl) and solvent B (90% methanol and 10 % 50 µM CH_3COONH_4). For ethylene thiourea, the mobile phase was isocratic, 95% water (pH adjusted to 3.0 with HCl, towards the low end of each column's pH limits of 2.5) and 5% methanol. A C18 5µ 250 mm × 4.6 mm (i.d.) PRISM RP column was used for analysis for this and the remainder of the pesticides. For 2,4-D, the mobile phase was composed of acetonitrile and water (55:45, pH was adjusted to 3 using phosphoric acid). For diazinon, the mobile phase was composed of acetonitrile and water (70:30, pH was adjusted to 3 using phosphoric acid). For carbaryl, the mobile phase was composed of acetonitrile and water (50:50, pH was adjusted to 3 by adding phosphoric acid). For analysis of carbofuran, carbaryl, bendiocarb, dioxacarb, and the mixture containing carbofuran and one or two of these carbamates, the mobile phase was composed of acetonitrile and water (40:60, pH adjusted to 3 using phosphoric acid). For analysis of promecarb, fenobucarb, and the mixture of carbofuran with one of these carbamates, the mobile phase was composed of acetonitrile and water (70:30, pH=3). Appropriate detection wavelengths were chosen for each pesticide. The concentration of hydrogen peroxide was analyzed by titration in acidic solution using potassium permanganate standard solution.

When degradation product identification was required, GC/MSD was enlisted. After 3 min of treatment by AFT under typical operating conditions, 15 mL of anodic solution was taken out and immediately extracted using 3 mL hexane. The sample was analyzed by an Agilent 6890N Network GC System equipped with an Agilent 5973 Network Mass Selective Detector and Agilent 7683 Series Injector. GC conditions have been reported *(21)*.

Results and Discussion

AFT was developed as an improvement to the electrochemical Fenton treatment (EFT) method. By separating the anode and cathode half cells, the Fenton reaction could occur at an optimum pH. Since water is being reduced at the cathode and forming hydroxide ions (eq. 3), effluent from the anode and cathode half cells can be mixed after the reaction, providing wastewater close to neutral pH.

The first pesticide degraded using AFT was ethylene thiourea (ETU) *(14)*. Degradation rates for ETU from the previous electrochemical treatment method (at neutral pH) were comparable to AFT with complete degradation of the parent compound in under 1.5 min. Classical Fenton degraded the same concentration of ETU in under 30 seconds. These fast reactions appeared to be zero order. Delivery rate of ferrous ion and $H_2O_2:Fe^{2+}$ were optimized in a series of experiments, and water soluble degradation products, ethylene thiourea and 2-imidazolin-2-yl sulfonic acid (Im-SO_3H), were tentatively identified via HPLC retention time matching as shown in Figure 2 *(14)*. In this work it was shown that AFT was at least comparable to the previous method and that it was competitive with the classical Fenton treatment.

AFT studies of atrazine degradation were carried out and compared to EFT and classical Fenton *(18)*. AFT degraded 135 µM atrazine rapidly to less than 1% of the initial concentration within three minutes, which is approximately ten times faster than under electrochemical Fenton treatment *(11)*. Classical Fenton treatment was reported to have degraded 98% of a similar concentration of

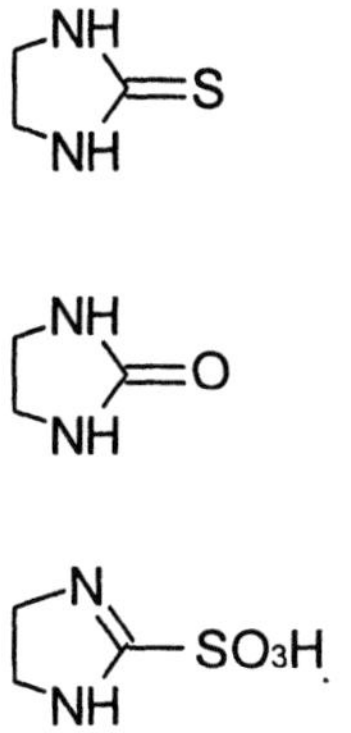

Ethylene thiourea, ETU

Ethylene urea, EU

2-Imidazolin-2-yl sulfonic acid, Im-SO_3H

Figure 2. Chemical structures of ethylene thiourea (ETU) and two degradation products, ethylene urea (EU) and 2-imidazolin-2-yl sulfonic acid (Im-SO_3H) (*14*).

atrazine in ≤30 seconds, with 1% remaining after 24 hours (7). Thus, the AFT compared well. Dealkylated and dechlorinated products of atrazine from AFT degradation were followed via HPLC, and it appears that dealkylation occurs before dechlorination. Figure 3 (18) shows a proposed degradation pathway for atrazine with AFT. The formation of the dealkylated, dechlorinated product, ammeline, is interesting because it is favorably formed under AFT conditions after only 10 minutes. In the classical Fenton work there was no quantifiable formation of this product after 11.5 hours. Dechlorination is often equated with detoxification, so these results were quite promising for AFT.

AFT Kinetic Model

An important advancement in use of the AFT method was the development of a kinetic delivery model to describe the degradation of a given pesticide (19). The model was demonstrated with the widely used herbicide, 2,4-D. This model accounts for the two important reactions occurring during treatment, the Fenton reaction and the reaction of the hydroxyl radical with the pesticide. The model can be summarized as follows. During AFT treatment, ferrous ion is constantly delivered into an anodic half-cell by electrolysis at a fixed rate. It is also continuously consumed by reaction with hydrogen peroxide, which is delivered simultaneously into the reaction system. We assume (i) the concentration of ferrous iron in the reaction system is constant; (ii) hydrogen peroxide can be accumulated in the reaction system when the ratio of hydrogen peroxide to ferrous ion is larger than 1; (iii) the Fenton reaction obeys second order kinetics; (iv) the instantaneous concentration of hydroxyl radical is proportional to its generation rate; (v) the kinetics of the hydroxyl radical reaction with organics obey second order. Based on these assumptions, the following equation can be established to describe the changes of pesticide concentration during AFT:

$$\ln \frac{[C]_t}{[C]_0} = -\frac{1}{2} K \lambda \pi \omega v_0^2 t^2 \qquad (4)$$

where, $K = kk_1$ ($\mu M^{-2} \cdot min^{-2}$), k ($\mu M^{-1} \cdot min^{-1}$) and k_1 ($\mu M^{-1} \cdot min^{-1}$) are the second order rate constants of the Fenton reaction and the reaction between hydroxyl radical and target compound, respectively; $[C]_0$ (μM) and $[C]_t$ (μM) are the concentrations of target compound at 0 and t min, respectively; λ (min) and π (min) are the average life of the hydroxyl radical and ferrous ion, respectively; ω is a constant related to the delivery ratio of hydrogen peroxide to ferrous ion and to the consumption ratio of hydrogen peroxide; v_0 ($\mu M \cdot min^{-1}$) is the delivery rate of ferrous ion by electrolysis; and t (min) is time. The fitting of the model to

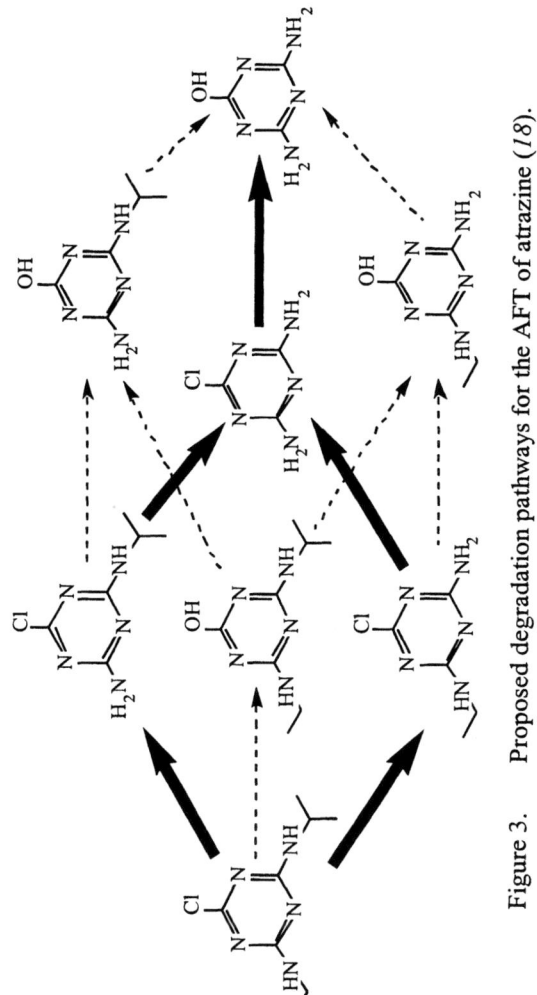

Figure 3. Proposed degradation pathways for the AFT of atrazine (*18*).

the experimental data for degradation of 2,4-D with different delivery rates of iron is shown in Figure 4 (*19*). This model provides a quantitative measure of rate that is extremely useful in many ways. It is a powerful tool for measuring hydroxyl radical reactions by allowing us to determine hydroxyl radical reaction rate constants and degradation rates for a variety of pesticides using competitive kinetics by measuring the change in concentration of only one probe pesticide, in this case, 2,4-D. Hydroxyl radical reaction rates measured in this manner were in excellent agreement with literature values. Since the model provides a rate parameter for a given pesticide that is specific to the AFT, degradation of the pesticide can be used as a probe to optimize the system and determine its efficiency under different conditions.

Diazinon was studied in order to determine its degradation pattern and also to explore the use of various electrolytes and their optimum concentrations for AFT. It was found that there was no difference in oxidation rate of diazinon when either NaCl or KCl was used as the electrolyte; the rate was slightly decreased when $NaSO_4$ was the electrolyte. When $NaNO_3$ was used as the electrolyte, there was a significant decrease in the oxidation rate, and the data could no longer be fitted by the AFT delivery model. It was postulated that nitrate is preferentially degraded over water at the cathode, changing the cathode potential and affecting reactions at the anode so that oxidation of water will occur preferentially over oxidation of iron. This would affect the assumptions of the AFT kinetic model and result in a poor fit of the data.

Electrolyte (NaCl) concentrations were varied in the anode and the cathode and the effect on the oxidation rate parameter for diazinon was measured (Figure 5 (*20*)). In addition, calculated current efficiencies that relate to the total amount of iron produced with the same variation in electrolyte concentrations in the half cells were also plotted. These results confirmed that an anode concentration of 0.04M NaCl and a cathode:anode electrolyte concentration ratio of 4:1 were optimum for degradation of diazinon.

The activation energy of the combined Fenton reaction and hydroxyl radical oxidation of diazinon was calculated by running the degradation experiments at several temperatures. The activation energy of diazinon oxidation by anodic Fenton treatment was estimated to be 12.6±0.6 KJ·mole^{-1}, which is less than the value in aqueous chlorine treatment. A relatively stable degradation product of diazinon is diazoxon, which appears and then is completely degraded after 3 minutes.

The AFT method also degrades the formulated product of diazinon, and the diazinon degradation can be fitted by the AFT kinetic model. The degradation is somewhat slower than that of the pure standard, due to competition for hydroxyl radicals by other compounds in the formulation.

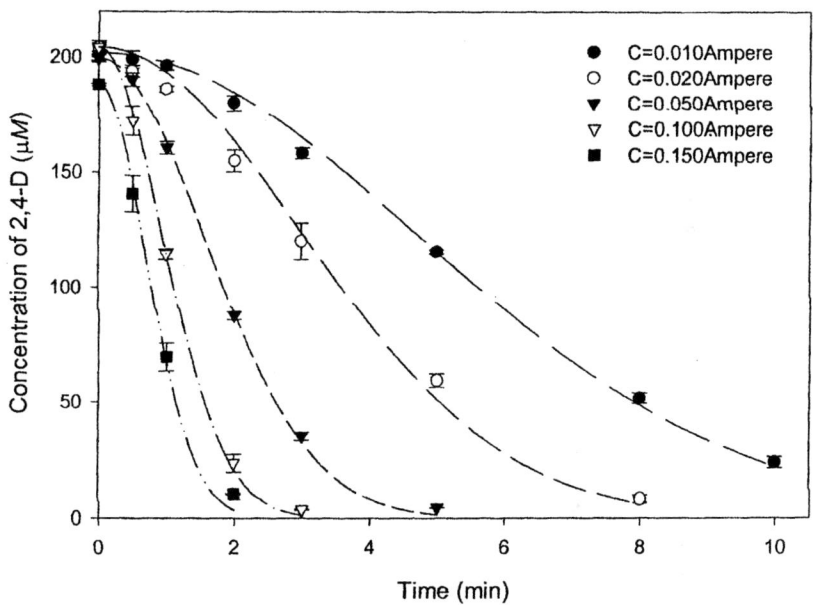

Figure 4. Degradation of 2,4-D (200μM) in with different delivery rates of Fenton reagent at a ratio of 10:1 H_2O_2:Fe^{2+}. Lines are AFT model fitting results
(Reproduced from reference 19. Copyright 2001 American Chemical Society.)

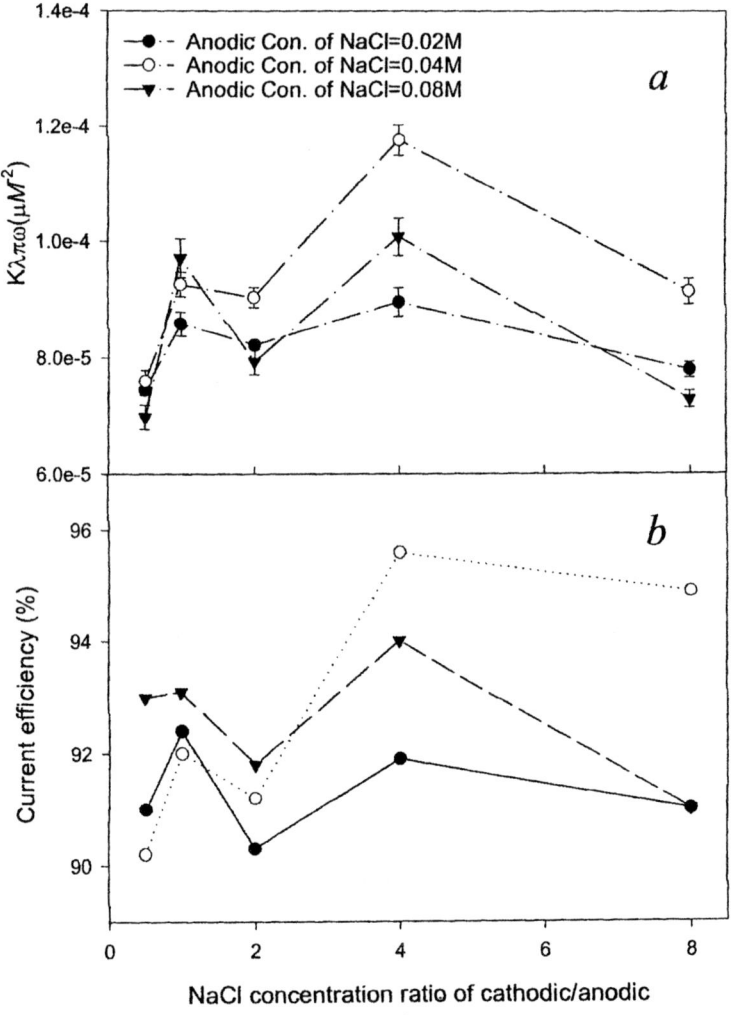

Figure 5. Variation of oxidation rate parameter (ln (Kλπω) for diazinon (a) and electrolysis current efficiency (b) with NaCl concentration ratio in the two half-cells
(Reproduced with permission from reference 20. Copyright 2002 Elsevier.)

Membrane AFT

A major advance in the development of the AFT was the use of an ion exchange membrane to replace the salt bridge. The apparatus is depicted in Figure 6 (*21*). The degradation of carbaryl was investigated at the same time as an ion exchange membrane was selected and electrolyte concentrations were optimized (*21*). Both cation and anion exchange membranes were investigated. The carbaryl degradation data were fitted by the AFT kinetic model and the rate parameter was used as a measure of treatment efficiency to evaluate the two membrane types and optimize electrolyte concentration. Results are shown in Figure 7 (*21*). One could attain higher treatment efficiency at lower electroyte concentration with the anion exchange membrane. Thus, because it is advisable to introduce the least amount of NaCl into the treatment system, the optimum treatment conditions for the experiments included the anion exchange membrane with an anodic NaCl concentration of $0.02M$ and a cathodic NaCl concentration of $0.08M$. It was also obvious from the efficiency data that a high cathode/anode electrolyte ratio is more favorable for the anion exchange membrane than the cathode exchange membrane. It was postulated that since the anion membrane only allows cathodic Cl^- and OH^- to pass through to the anodic half-cell, the increased NaCl concentration ratio (cathode/anode) is beneficial for Cl^- to compete with OH^- in movement across the membrane, exerting little effect on the anodic pH. Thus, the treatment efficiency should increase with the increase of NaCl concentration ratio (cathode/anode) in the anion membrane system.

In contrast, in the cation membrane AFT system, the membrane only allows anodic Na^+ and H^+ to pass through to the cathodic half-cell. With an increase of cathodic NaCl concentration, the movement of Na^+ from anode to cathode may become more difficult because of the concentration gradient between the two half-cells. More H^+ than Na^+ may move into the cathodic half-cell. To regenerate H^+ and keep the optimal pH for the Fenton reaction, some of the Fenton reagent may be consumed. This can result in a decrease of treatment efficiency in the cation membrane AFT system when the NaCl concentration ratio (cathode/anode) is high. The functional stability of the membrane was also evaluated in this study, and it was found that treatment efficiency for degradation of carbaryl did not change after the membrane was used 100 times.

Competitive Kinetics

The AFT treatment system and the AFT kinetic model were successfully employed to study the competitive kinetics, the degradation products, and proposed degradation mechanisms of six carbamate pesticides (*22*) A complete study of the degradation of carbofuran had already been done (*23*). Based on

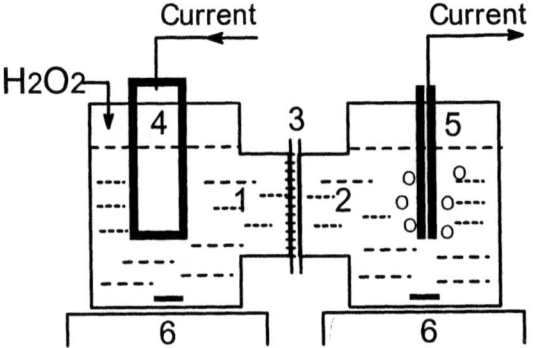

Figure 6. Membrane anodic Fenton treatment appartus. (1) Anodic half-cell; (2) Cathodic half-cell; (3) Ion exchange membrane; (4) Iron plate; (5) Graphite stick; (6) Magnetic stirring plate
(Reproduced from reference 21. Copyright 2002 American Chemical Society.)

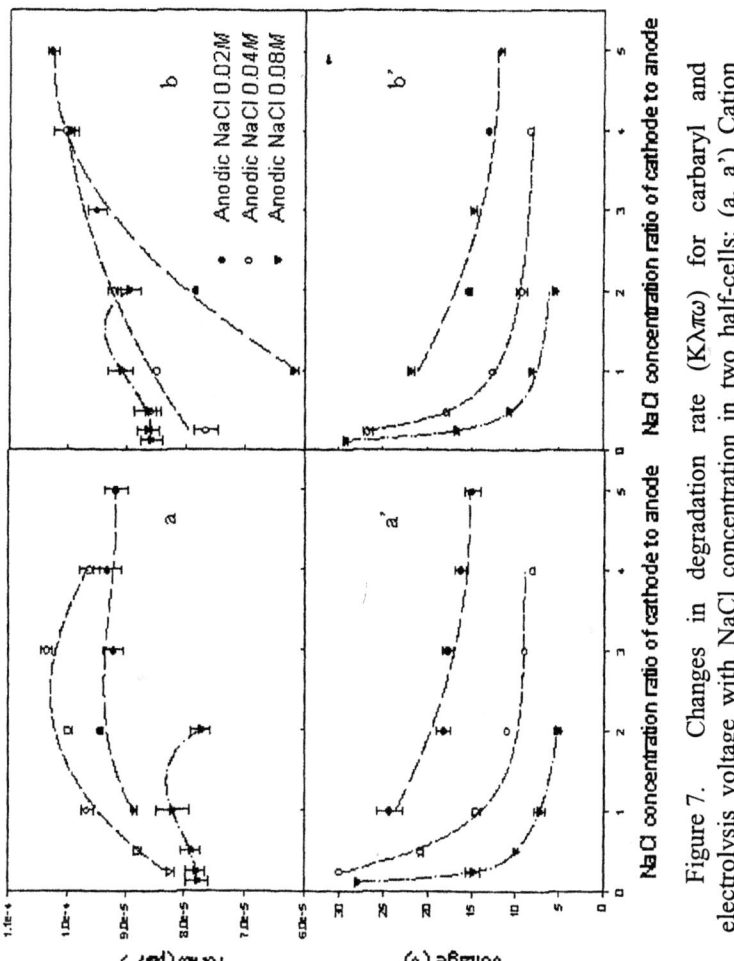

Figure 7. Changes in degradation rate (K$\lambda\pi\omega$) for carbaryl and electrolysis voltage with NaCl concentration in two half-cells: (a, a') Cation exchange membrane AFT; (b, b') Anion exchange membrane AFT *(Reproduced from reference 21. Copyright 2002 American Chemical Society.)*

this work, the competitive degradation between carbofuran and a coexisting carbamate pesticide in the same solution was studied using the AFT kinetic model. This work was repeated for five carbamate pesticides: dioxacarb, carbaryl, fenobucarb, promecarb, and bendiocarb. The degradation data were determined by analysis of each pesticide and were fitted by the AFT kinetic model. Since λ, π, ω, and v_0 should be the same for each of two pesticides in a given solution, and the Fenton reaction rate constant is also the same, the difference in degradation rate can be determined by k', the hydroxyl radical reaction rate constant for a given pesticide. Thus, the degradation rates of of the two pesticides are proportional to their hydroxyl radical reaction rate constants. The following equation can describe the ratio of coexisting carbamates.

$$\frac{k'_1}{k'_2} = \frac{(K\lambda\pi\omega)_1}{(K\lambda\pi\omega)_2} \tag{5}$$

Using the literature value for carbofuran, hydroxyl radical reaction rate constants were calculated for the other five pesticides (Method II).

An indirect method (Method I) was also employed to determine these hydroxyl radical reaction rate values. Since a relationship has been developed between the degradation rate of carbofuran and its initial concentration in the previous work,

$$\ln(K\lambda\pi\omega) = -4.478 - 1.062\ln C_{initial} \tag{6}$$

it was possible to measure only the concentration of carbofuran and to determine the degradation rates of a given coexisting pesticide from the change in degradation rate of the carbofuran. This method was demonstrated in our previous work with 2,4-D. Hydroxyl radical reaction rates were determined by both methods and an excellent correlation between the methods was demonstrated (Figure 8).

Biodegradability

From the identification of the degradation products by GC-MS, the degradation of the six carbamate pesticides demonstrated attack at the carbamate branch by hydroxyl radicals. In addition, for promecarb and carbaryl, hydroxyl radicals can also initiate attack at the alkyl site. These degradative steps result in substituted phenols, which may or may not be the final products from AFT treatment. It can be demonstrated by measuring the BOD_5/COD of a solution of all six carbamates that the biodegradability of the solution is significantly

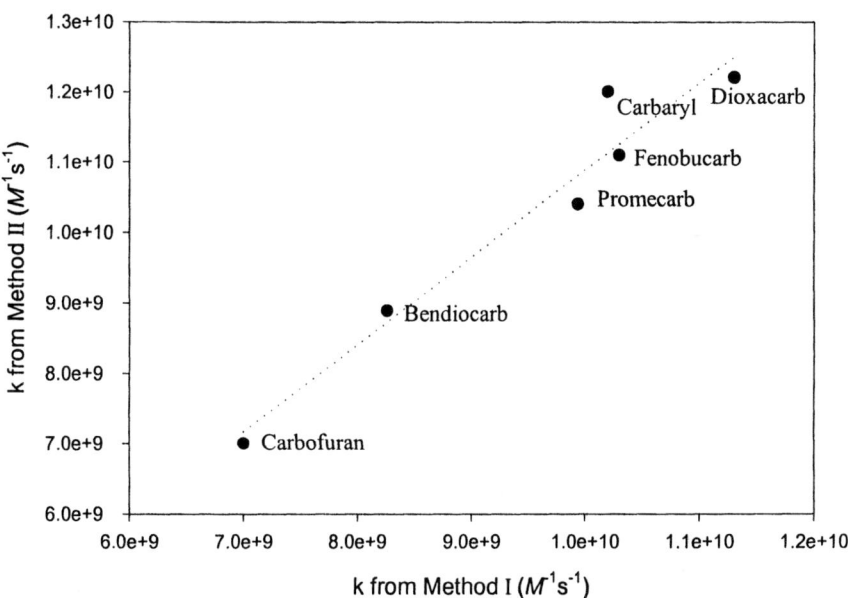

Figure 8. Hydroxyl radical reaction rate constants calculated from the direct method (Method II) plotted against the rate constants calculated from the indirect method (Method I) for six carbamate pesticides.

enhanced after 15 minutes of treatment since it goes well above the accepted value of 0.3 that indicates biodegradability. The initial value before AFT treatment is 0.04, at 3 minutes it is 0.34, and after 15 minutes it is 0.71. These results, combined with some earthworm studies that show decreased toxicity of the solution after 10 minutes of AFT, indicate that AFT is a very promising pretreatment for combination with microbial degradation of pesticides. Preliminary work in the laboratory with metribuzin also shows that the AFT effluent can be used safely with a microbial consortium after the addition of catalase to degrade excess hydrogen peroxide.

Conclusions

Anodic Fenton treatment has been shown to be an effective advanced oxidation method for degradation of pesticides. It has some advantages over classic Fenton treatment in that it produces more extensive degradation of toxic products. Improvements to AFT, such as membrane separation of the half-cells, have made it a more practical method to be scaled up for field use. The AFT kinetic model has been shown to be extremely useful for optimizing the system and for study of competitive kinetics of coexisting pesticides. AFT has also been shown to improve the biodegradability of refractive compounds and to decrease the toxicity of a mixture of pesticides. It holds significant promise for the future.

References

1. Kolpin, D.W.; J.E. Barbash; R.J. Gilliom, *Environ. Sci. Tech.* **1998** *32*, 558-565.
2. *Pesticide Waste Management. Technology and Regulation;* Bourke, J.B.; Felsot, A.S.; Gilding, T.J.; Jensen, J.K.; Seiber, J.N., Eds.; American Chemical Society, Washington, DC, 1992.
3. Felsot, A.F. *J. Environ. Sci. Health* **1996** *B31*, 365-381.
4. Sedlack, D.L.; Andren, A.W.. *Environ. Sci. Technol.* **1991** *25*, 777-782.
5. Sedlak, D.L.; Andren, A.W.. *Environ. Sci. Techno.* **1991** *25*, 1419-1427.
6. Haag, W.R.; Yao, C.C.D. *Environ. Sci. Technol.* **1992** *26*, 1005-1013.
7. Arnold, S.M.; Hickey, W.J.; Harris, R.F. *Environ. Sci. Technol.* **1995** *29*, 2083-2089.
8. Arnold, S.M.; Hickey, W.J.; Harris, R.F.; Talaat, R.E. *Environ. Toxicol. Chem.* **1996** *15*, 1255-1262.
9. *Proceedings of Annual Meeting of Air & Waste Management Association;* Chen, C.T.; Tafuri, A.N; San Antonio, TX. 1995.

10. Tang, W.Z.; Huang, C.P. *Environmental Technology* **1996** *17*, 1371-1378.
11. Pratap, K.; Lemley, A.T. *J. Ag. and Food Chem.* **1998** *46*, 3285-3291.
12. Huston, P.J; Pignatello, J.J. *Wat. Res.* **1999** *33*, 1238-1246.
13. Bier, E.L.; Singh, J.; Li, Z.; Comfort, S.D.; Shea, P.J.. *Environ. Toxicol. Chem.* **1999** *18*, 1078-1084.
14. Saltmiras, D.A.; Lemley, A.T. *J. Ag and Food Chem.* **2000** *48*, 6149-6157.
15. Dowling, K.C.; Lemley, A.T. *J. Environ. Sci. Health* **1995** *B30*, 585-604.
16. Roe, B.A; Lemley, A.T. *J. Environ. Sci. Health* **1997** *B32*, 261-281.
17. Saltmiras, D.A.; Lemley, A.T. *J. Environ. Sci. Health* **2001** A36, 261-274.
18. Saltmiras, D.A.; Lemley, A.T. *Wat. Res.* **2002** *36*, 5113-5119.
19. Wang, Q.; Lemley, A.T. *Environ. Sci. Technol.* **2001** *35*, 4509-4514.
20. Wang, Q.; Lemley, A.T. *Wat. Res.* **2002** *36*, 3237-3244.
21. Wang, Q.; Lemley, A.T. *J. Ag. Food Chem.* **2002** *50*, 2331-2337.
22. Wang, Q.; Lemey, A.T. *Environ. Sci. Technol.* (under review).
23. Wang, Q.; Lemley, A.T. *J. Haz. Mat.* (in press).

Chapter 7

Chemical Detoxifying Neutralization of *ortho*-Phthalaldehyde: Seeking the "Greenest"

Peter C. Zhu[1], Charles G. Roberts[1], and Jiejun Wu[2]

[1]Advanced Sterilization Products, Johnson & Johnson, 33 Technology, Irvine, CA 92618
[2]J&J Pharmaceutical Research and Development, LLC, 3210 Merryfield Row, San Diego, CA 92121

CIDEX® OPA Solution (containing 0.55% *ortho*-phthalaldehyde, OPA) is the preferred choice over glutaraldehyde (GTA) as a high-level disinfectant for hospital instrument processing due to its low volatility and high stability. Although OPA degrades naturally in the environment, residual OPA in discharge water may be toxic to organisms in the receiving water, and these solutions should ideally be detoxified prior to discharge. Even though OPA can be neutralized by sodium bisulfite, the neutralized solution is toxic to fish. OPA can also be scavenged by passing through solid supports with covalently bonded amines. However, this method seems not very practical. Ammonia precipitates OPA effectively and instantly from its water solution but the product mixture is toxic to fish. When OPA is neutralized by glycine, forming an Schiffs base (fish safe), the neutralized solution appears dark. New neutralization methods seem essential. NaOH (Cannizzaro reaction) neutralizes OPA. The end product shows no color change and is fish-safe. Hydrogen peroxide also neutralizes CIDEX® OPA Solution almost instantly at room temperature without color change, and the neutralized solution passes the fish test. Bleach seems as good as hydrogen peroxide but the end product fails pass the fish test. Detailed neutralization mechanisms are discussed.

© 2004 American Chemical Society

Introduction

CIDEX® OPA Solution (containing 0.55% *ortho*-phthalaldehyde, OPA, **1**) has become the preferred choice over glutaraldehyde (GTA, **2**) as a high-level disinfectant for hospital instrument processing.

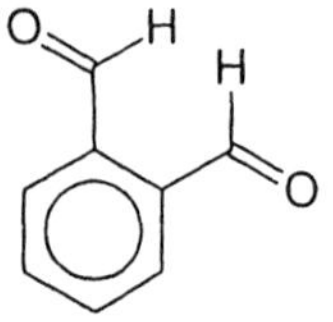

1
ortho-Phthalaldehyde
(OPA)

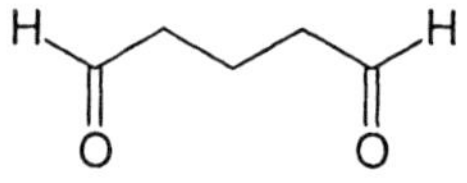

2
Glutaraldehyde
(GTA)

The advantages include low volatility (thus low toxicity to hospital personnel) and high stability (long shelf-life and thus less disposal due to expiration and hence more environmentally friendly). Although OPA degrades naturally and the disposal of CIDEX® OPA Solution after instrument processing is not regulated in most of the states in US or in most other countries, it is regulated in the State of California and in Japan. With the ever-increasing use of OPA in hospitals worldwide, finding an ideal method to neutralize or detoxify it conveniently becomes an important research effort.

OPA can be neutralized with sodium bisulfite as shown (Fig. 1). The neutralization is very fast and a colorless aldehyde-bisulfite adduct, **3**, is formed. However, the product is toxic to fish even though sodium bisulfite itself is not.

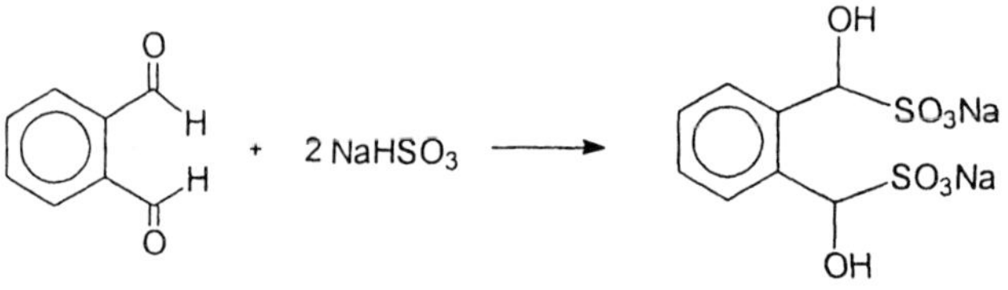

OPA + 2 NaHSO₃ → **3** Colorless, Toxic to fish

*Figure 1. OPA neutralization by sodium bisulfite. The end product, **3**, is toxic to fish.*

Like neutralization of glutaraldehyde,[1a] glycine neutralizes OPA, forming a fish-safe Schiff's base, **4**.[1b] However, the neutralized solution is dark. Zhu et al developed a method to reduce the dark Schiff's base (by NaBH$_4$ reduction) so that the final solution containing **5** shows no dark color[2a] as shown in Fig. 2.

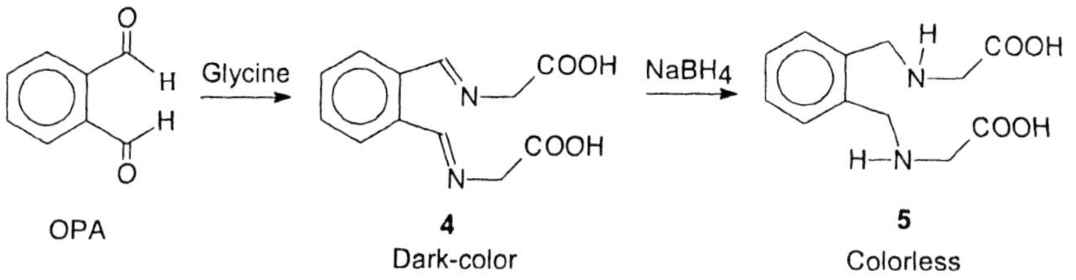

Figure 2. OPA neutralization by glycine followed by sodium borohydride reduction. The dark color of OPA-glycine adduct, 4, disappeared instantly. The disadvantage of this method is the use of a harmful reagent, sodium borohydride

Since this method involved the use of a reducing agent, sodium borohydride, Zhu developed several methods in which amines, such as **7**, were covalently linked to solid supports and water solution containing OPA was passed through the support. OPA was scavenged by forming a Schiff's base, **9**, with the amines on solid supports. Silica was a common choice for this solid support as shown Fig. 3.[2b]

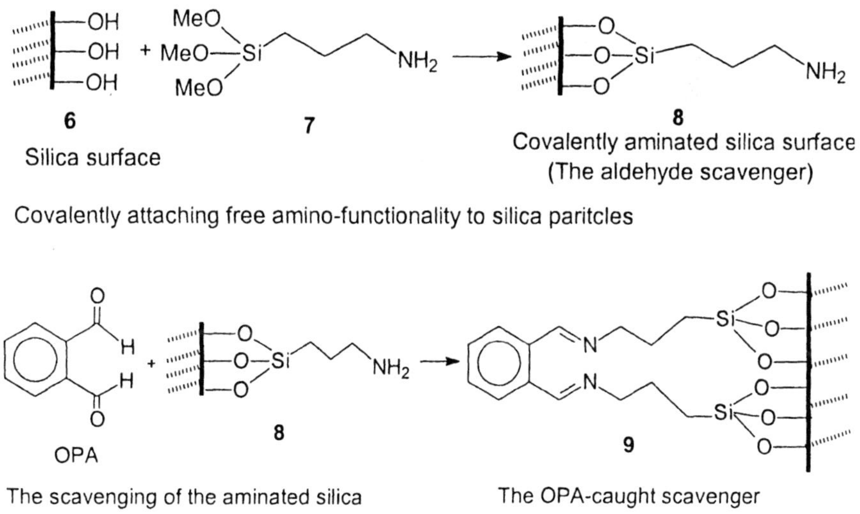

Figure 3. OPA neutralization by scavenging on a silica support. This method requires covalent binding of amines to a solid support and subsequent disposal of the solid support after neutralization.

Alternatively, cheap and biodegradable solid supports, such as starch-type supports 10, can be used (Fig. 4). However, this scavenging approach requires the subsequent disposal of a solid waste.

Interestingly, we found that household ammonia (diluted ammonia water solution) precipitates OPA effectively and instantly from its water solution at room temperature. However, the product mixture is toxic to fish and is not an ideal neutralization method for hospital users since qualified personnel and a filtration procedure (to remove toxic precipitate) are required. In some situations, this process may be acceptable.

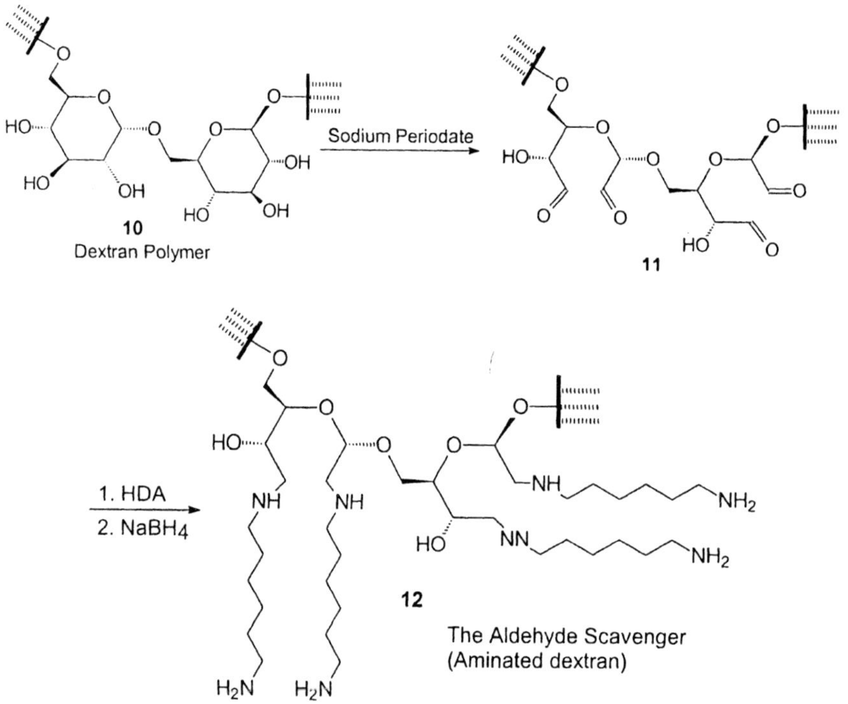

Figure 4. OPA neutralization by scavenging on dextran, 10.
A cheaper and biodegradable biopolymer like starch could be employed.
Still, this method requires covalent binding of amines to solid supports
and subsequent disposal of the solid support after neutralization.

Thus, an ideal method for OPA neutralization would be one that does not transform OPA to another toxic agent, one that does not need a solid support, and one that does not use an inconvenient agent, such as a reducing agent, and one that does not produce a dark colored solution. Our efforts in developing such an ideal, "green," detoxifying neutralization method are summarized.

Results and Discussion

In terms of oxidation status, aldehydes are located between alcohols and carboxylic acids. An aldehyde can be oxidized to a carboxylic acid or reduced to an alcohol. The general organic or biological reactivity of an aldehyde to participate organic reactions is high due to an aldehyde's easy reaction with amino residues in peptides and proteins. Thus, a logical way to neutralize an aldehyde is to consider reduction and oxidation reactions (Redox reactions) or disproportionation reactions. Since organic structure/biotoxicity relationships are not well established for aromatic aldehydes and because a Dakin reaction would lead to the formation of phenol-type compounds, great caution is needed for the handling the neutralization products considering their toxicity. To ensure finding a good neutralization method, it was essential (1) to be precautious for compounds that might form after neutralization; (2) to conduct fish survival tests by the standard procedures;[5] and (3) to chemically analyze the neutralization solution, especially if a solution passes the fish tests. All the possible products formed by Redox or disproportionation approaches were investigated before the real "wet chemistry" experimentation (see Fig. 5, some known ID50 data are shown). As can be seen from Fig. 5, the very first compound to avoid is catechol, **17**, due to its corrosiveness to the mucous membranes. The next compound to avoid is 2-hydroxybenzaldehyde, **16**, due to its possible effects on fertility, embryos or fetuses. On the other hand, compounds such as **19** (phthalic acid), **20** (salicyclic acid) and **15** (phthalide) are the desired final products due to their low toxicity. LD50 and other safety data are not available for **13** and **21**.

Two safe neutralization methods of the above type have been developed. The first is the Cannizzaro reaction using sodium hydroxide at room temperature. The second, perhaps the more preferred method, is an oxidation reaction employing diluted hydrogen peroxide.

Neutralization with Cannizzaro Reaction by Sodium Hydroxide

One of the two major "preferred" OPA neutralization methods is the Cannizzaro reaction (a disproportionation reaction) catalyzed by NaOH at room temperature, done by simply mixing OPA solution with 50% NaOH at room temperature.[2d] The resulting solution showed no color changes or toxicity to fish after pH adjustment to 7.[2d] Since a Cannizzaro reaction of dialdehydes is rarely studied, three possible routes were examined as shown in Fig. 6.

From chemical reaction evidence (color development with glycine), we first proposed that 2-carboxybenzaldehyde, **18**, and 2-hydroxymethylbenzaldehyde, **21**, were formed from an intermolecular Cannizzaro reaction *(Route 3)* as shown in Fig. 6., since the following step-wise mechanism seemed reasonable (Fig. 7).

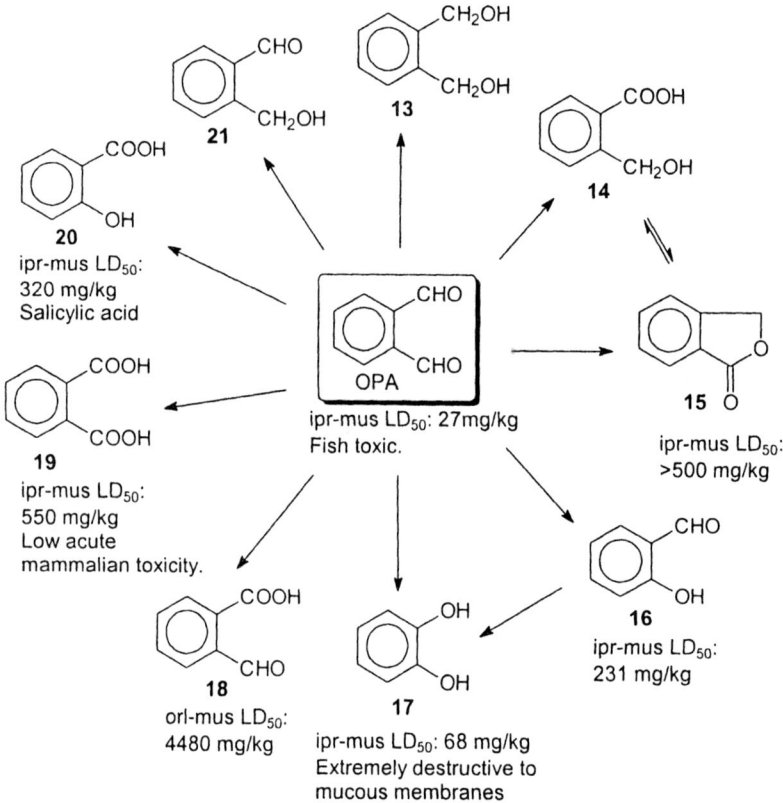

Figure 5. Possible products formed upon OPA neutralization by redox or disproportionation reactions.

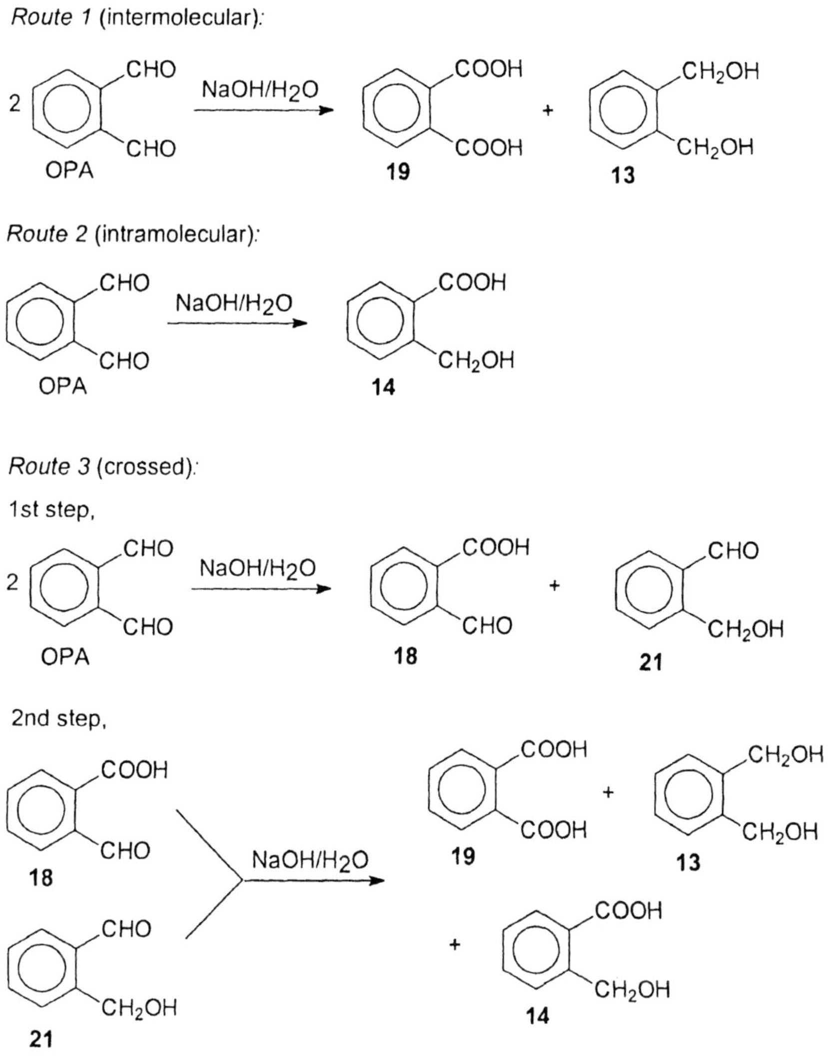

Route 1 (intermolecular):

2 OPA (CHO, CHO) →[NaOH/H₂O] **19** (COOH, COOH) + **13** (CH₂OH, CH₂OH)

Route 2 (intramolecular):

OPA (CHO, CHO) →[NaOH/H₂O] **14** (COOH, CH₂OH)

Route 3 (crossed):

1st step,

2 OPA (CHO, CHO) →[NaOH/H₂O] **18** (COOH, CHO) + **21** (CHO, CH₂OH)

2nd step,

18 (COOH, CHO) + **21** (CHO, CH₂OH) →[NaOH/H₂O] **19** (COOH, COOH) + **13** (CH₂OH, CH₂OH) + **14** (COOH, CH₂OH)

Figure 6. OPA neutralization by Cannizzaro reactions: Three possible routes may lead to different end products.

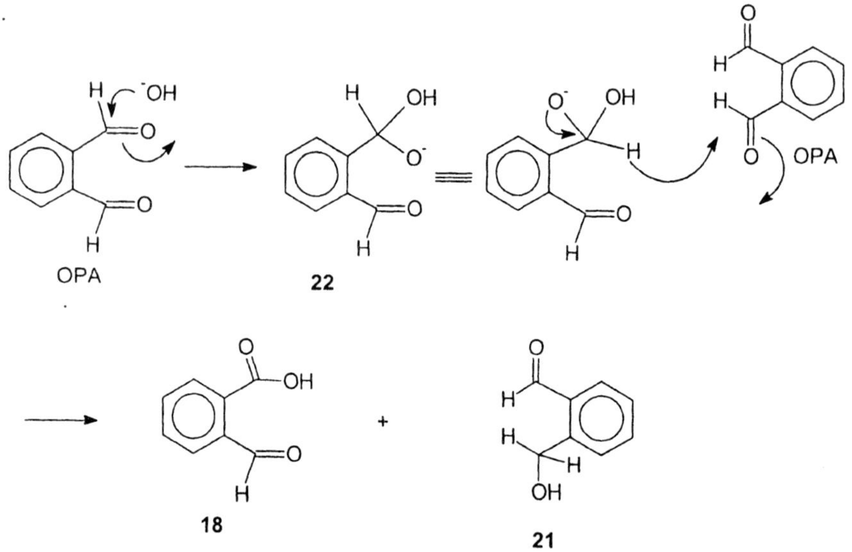

Figure 7. OPA neutralization by Cannizzaro reactions: A stepwise mechanism postulated based on the color reaction between the neutralized solution and glycine.

However, this is not what happened. HPLC/MS reveals that only 2-(hydroxymethyl)benzoic acid, **14**, is formed when the product is not acidified (electron spray, negative mode), or phthalide, **15**, is formed when the product is acidified or when the pH is adjusted to neutral (electron spray, positive mode). This is a desired result since 2-hydroxymethylbenzaldehyde, **21**, as discussed above, was not a preferred by-product. Thus *Route 1* in Fig. 6 was confirmed to be the actual neutralization path. The corresponding stepwise mechanism was proposed as follows (Fig. 8). Thus, the intramolecular hydride transfer of **22** as shown in Fig. 8 is more favorable than that of an intermolecular process shown in Fig. 7.

This was further confirmed by GC/MS analysis by acidifying the NaOH neutralized OPA solution to pH3 followed by immediate extraction with ethyl acetate. GC/MS shows that phthalide is the only peak present in the ethyl acetate solution. This easy ring closure is not surprising considering the fast kinetics of ring formation and the stability of the final product (Fig. 9).

The advantages of this method include the neutralization at room temperature and the lack of fish toxicity of the final product (after pH adjustment from basic to neutral). However, compared to hydrogen peroxide or bleach approaches described below (both instant neutralization), the neutralization took a much longer time (1-2 days). Also, this method involved the use of a corrosive agent; sodium hydroxide and preferred subsequent pH adjustment before disposal.

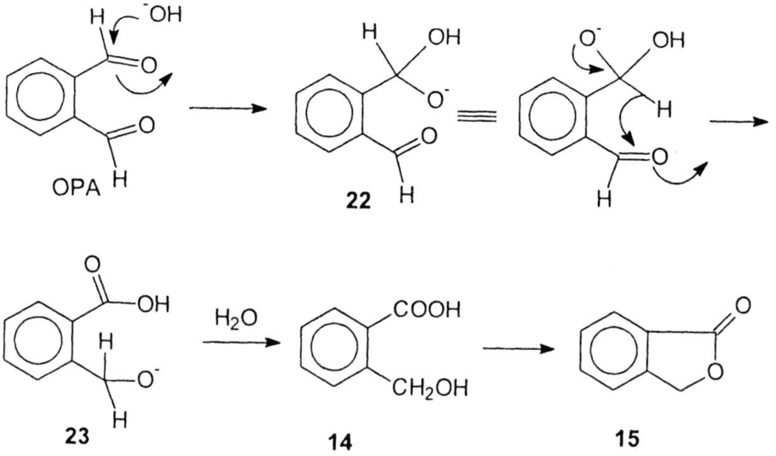

Figure 8. OPA neutralization by Cannizzaro reactions: mechanism proposed based on HPLC/MS results.

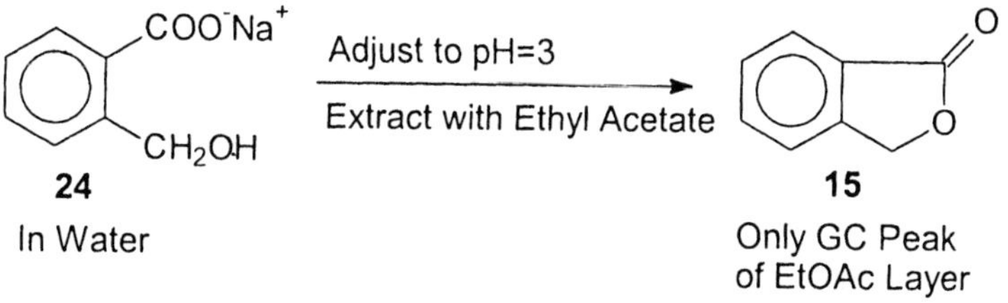

*Figure 9. OPA neutralization by Cannizzaro reactions: final product **15** independently confirmed by GC/MS.*

Neutralization with Hydrogen Peroxide by Oxidation

The other, preferred new method is hydrogen peroxide oxidation.[2c] Hydrogen peroxide (at different concentrations) neutralizes CIDEX® OPA Solution almost instantly at room temperature (based on color reaction of OPA with glycine). No color changes are detectable after neutralization. The neutralized solution passes the fish test[2c, 3] (this becomes an important criterion if the neutralized solution goes to rivers or oceans).

The oxidation of aromatic aldehydes could be complicated. In addition to direct oxidation to the dicarboxylic acid (phthalic acid), an aromatic aldehyde carbonyl could also be oxidized and then rearranged and hydrated to form a phenol via aryl formates (Dakin reaction).[4] Thus, when treating *p*-hydroxybenzaldehyde, **25**, with basic hydrogen peroxide, the following mechanism is actually involved (Fig. 10).

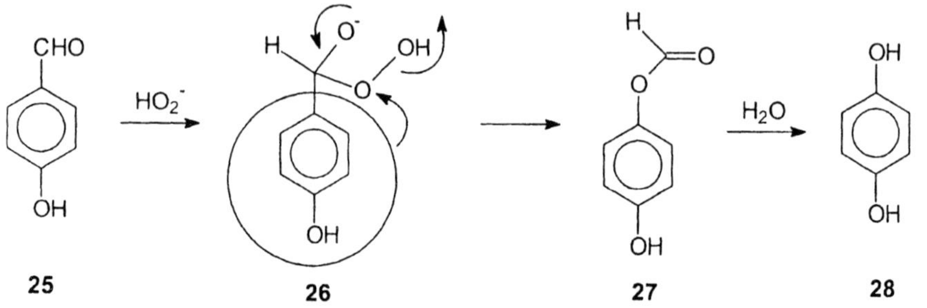

Figure 10. An example of the Dakin reaction: an aromatic aldehyde is converted to a phenol.

Assuming both direct oxidation and Dakin reaction are possible during OPA neutralization by hydrogen peroxide, three possible products, **19**, **20** and **17**, could form (Fig. 11). One of them, catechol, is very toxic.

What product will actually form? Will catechol form via the Dakin reaction? Aromatic aldehydes in Dakin reactions usually need to have an electron-donating group, like hydroxyl, alkoxyl or amino groups, in the *ortho-* or *para-* positions. This is illustrated in a few more examples shown below (Fig. 12).[5]

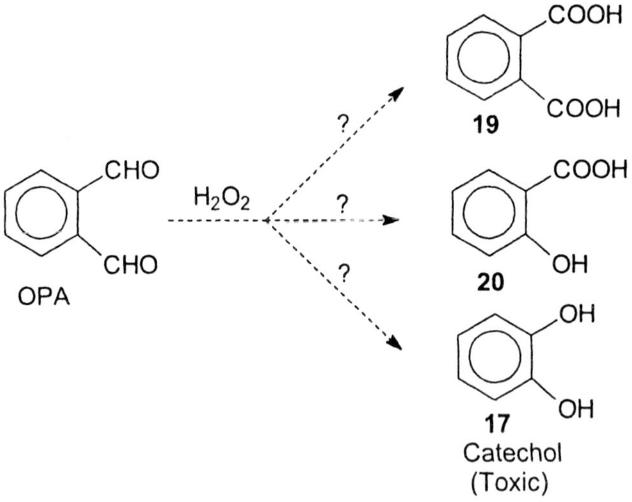

Figure 11. OPA neutralization by hydrogen peroxide: will catechol form?

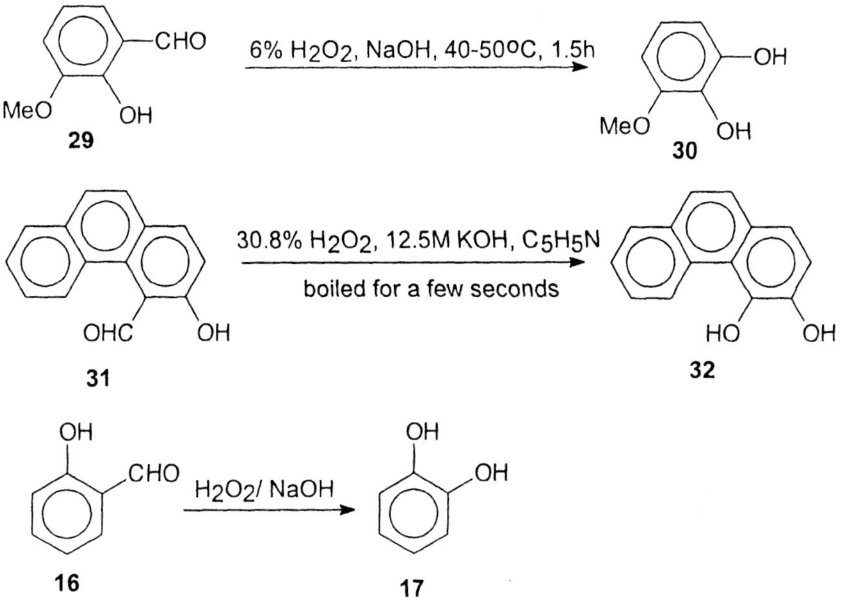

*Figure 12. A closer look at Dakin reactions: presence of **ortho** electron-donating groups is usually needed.*

The situation is different in the case of OPA. The two aldehyde carbonyls in OPA are electron withdrawing to each other. Thus, the Dakin reaction seems unlikely to occur during OPA neutralization with hydrogen peroxide. More specifically, we expect that the first step Dakin reaction forming **16** (Fig. 13) with not occur. If the first step (the more difficult step) did occur, the second step could follow since it seemed an easier step (forming the highly toxic catechol as the main product. This is highly undesirable).

Surprisingly, the oxidation of OPA by hydrogen peroxide without adding a base goes on unexpectedly rapidly and smoothly. The normal Dakin reaction usually requires either heating or prolonged reaction time, causing speculation that OPA may undergo some other reaction. With caution, we propose that the phthalic acid is the main oxidation product based on color changes.[2c] This proposal is confirmed by HPLC/MS results that show phthalic acid to be the only product (electron spray, negative and positive ionization). The other two possible products are not observed (Fig. 13). This result is desirable and interesting. The ease of oxidation by hydrogen peroxide is not expected for common aromatic aldehydes.

When bleach (5% NaOCl) is used, the reaction rate appears to be as fast as that using hydrogen peroxide at room temperature. No color change occurs. However, the neutralized solution is toxic to fish. The chemistry and the cause of fish toxicity are still being investigated.

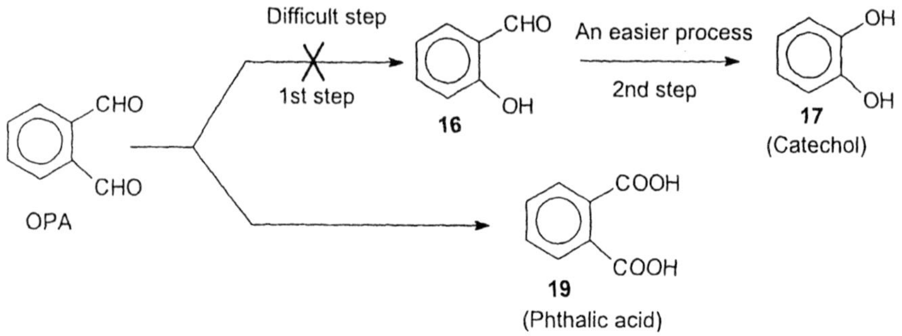

Figure 13. Ideal OPA neutralization by hydrogen peroxide: phthalic acid might be the only product.

Interestingly, many other strong oxidants, such as 5% potassium persulfate and 5% sodium chlorite also neutralize OPA but at much slower reaction rates.

Conclusion

Many chemical neutralization methods have been attempted. The hydrogen peroxide oxidation and sodium hydroxide catalyzed Cannizzaro reaction are the two preferred methods for the OPA detoxifying neutralization. Further, the diluted hydrogen peroxide approach, using ~3% hydrogen peroxide might be the "greenest" method. All it takes is the simple mixing of CIDEX® OPA Solution with diluted hydrogen peroxide at room temperature for a couple of minutes. Any excess hydrogen peroxide would be harmless since it is benign and decomposes to water in the environment.

A systematic original research of a wide range of possibilities identified two methods, which could be used safely for OPA neutralization, both meeting the needs of environmental protection and customer convenience around the world.[2c, 2d] The hydrogen peroxide approach meets the "greenest" demand possible. Although this work focuses on the neutralization of OPA in its 0.55% solution, with slightly modification, solid OPA (or OPA solutions with other concentrations) or compounds with similar structures should be also applicable.

References

1. (a). Nagata, S., JP7204661, (b). Chen, X. and Roberts, C., US Application 09/321,964.
2. (a). Zhu, P., Chen, X. and Roberts, C., US Application 09/747,230. (b). Zhu, P., Chen, X. and Roberts, C., US Application 09/746,344. (c).

Zhu, P., US Application 09/896,461. (d). Zhu, P.; Roberts, C. and Chan-Myers, H., US Application 09/896,589.
3. CAL. CODE REGS., TIT. 22, 66696.
4. Dakin, H. D., *Amer. Chem. J.* **42**, 4177, **1909**
5. (a). Belli, A. and Giordano, C., *Synthesis* **15**, 477, **1980**; (b). Kaeding, W. W. and Collins, G. R., *J. Org. Chem.* **30**, 3750, **1965**; (c). Kabalka, G.W.; Reddy, N.K.; Narayana, C. Tetrahedron Lett. 1992; 33:865.

Chapter 8

Electrochemical Destruction of Triclosan

James Farrell, Jiankang Wang, and Ronald LeBlanc

Department of Chemical and Environmental Engineering, University of Arizona, Tucson, AZ 85721

Abstract: This research investigated electrochemical oxidation of triclosan using Ebonex® and boron-doped diamond (BDD) film anodes. Oxidative destruction of triclosan was conducted in both high pH aqueous solutions and in ethanol. At current densities of 5 mA/cm^2 and above, oxidation of triclosan in water was rapid due to the action of hydroxyl radicals generated from water oxidation. However, at current densities below 5 mA/cm^2, the electrodes in aqueous solutions were rapidly inactivated by a film of polymerized byproducts. Oxidation of triclosan in ethanol solutions was much slower than in water. The primary mechanism of triclosan oxidation in ethanol was indirect, and involved ethoxy radicals produced from ethanol oxidation. Product analysis showed that breaking the ether linkage was easier than opening the aromatic rings. Microtox® tests showed that residual triclosan was the major source of toxicity in the treated wastewater, despite byproduct concentrations that were significantly higher than triclosan concentrations.

Triclosan, shown in Figure 1, is a polychlorinated diphenyl ether and is similar in structure to common herbicides. Triclosan has been in use for more than 30 years as a broad spectrum antibacterial agent due to its ability to block one step in bacterial fatty acid synthesis (*1*). In addition to its high toxicity to bacteria, triclosan is also acutely toxic to fish, protozoa, and other aquatic organisms (*2*).

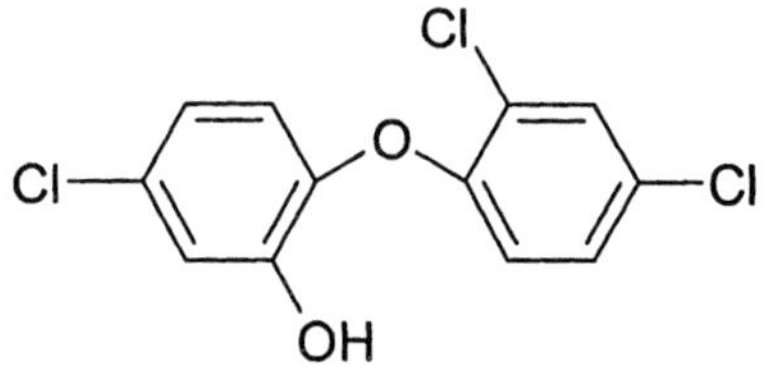

Figure 1. Triclosan (5-chloro-2-(2,4-dichlorophenoxy)-phenol).

Triclosan is commonly used in a wide variety of household products, including: soaps, toothpastes, deodorants, cosmetics, disinfectant solutions, footwear, and plastics. A recent study by the United States Geological Survey found that triclosan was the third most ubiquitous manmade compound in surface streams in the United States (*3*). It has been found entering wastewater treatment plants at concentrations on the order of 1 µg/L, and is only partially degraded in activated sludge treatment systems (*2, 4*). During wastewater treatment the phenolic group may become methylated, which increases the stability of the compound to photodegradation (*2*). Due to its environmental persistence, it has been found worldwide in lakes, sediments, aquatic organisms and human milk (*4*).

Aging stockpiles of triclosan often become contaminated by polychlorinated dioxins as a result of exposure to heat or ultraviolet light. Because chlorodioxins are among the most highly carcinogenic compounds known, stockpiles adulterated by even trace levels of these compounds cannot be used. Incineration of contaminated stockpiles on a small scale is impractical due to the precautions that must be taken to avoid producing chlorinated organic byproducts during the incineration process. The high toxicity of triclosan to bacteria precludes disposal of contaminated or aging stockpiles via biodegradation.

In addition to the disposal problems associated with powdered triclosan, there are also disposal problems associated with triclosan containing organic solvents. Solutions containing up to 2.5% triclosan in ethanol or isopropanol are commonly used for disinfecting medical equipment. Because of the high toxicity of triclosan to bacteria, and the recalcitrance of triclosan in activated sludge treatment systems, disposal of these solutions into sanitary sewers may result in process disruptions and eventual release of triclosan into the aquatic environment. These disposal limitations were the motivation for investigating the electrochemical destruction of triclosan. The goal of the electrochemical oxidation scheme was to convert triclosan into biodegradable compounds that could be treated in conventional activated sludge treatment plants.

Background

Electrochemical Oxidation

Commonly used advanced oxidation processes for removing organic contaminants from water generally rely on hydrogen peroxide as the oxidizing agent. Although hydrogen peroxide is stable in water, it may be decomposed to highly reactive hydroxyl radicals. The distinguishing feature between different advanced oxidation processes is the method of generating hydroxyl radicals from the H_2O_2 reagent. Ultraviolet light, ozone and Fe^{2+}/Fe^{3+} are often used to decompose the H_2O_2. Each of these methods has its advantages and disadvantages, and the most appropriate method usually depends on the pH and the composition of the wastewater.

Although electrochemical oxidation of organic compounds in wastewater is not widely utilized in commercial applications, it has the potential to be more effective and less costly than the three most commonly used methods. Lower costs will result from the fact that electrochemical oxidation does not require the addition of the H_2O_2 reagent, which itself is often electrochemically synthesized. Therefore, electrochemical oxidation will result in a simpler process and save the expenses associated with concentrating, transporting, and storing the H_2O_2 reagent. Also, elimination of the need for UV light, ozone or Fe^{2+}/Fe^{3+} will save additional operating expenses.

Electrochemical oxidation of organic compounds is generally performed using noble metal or metal oxide anodes. In dilute aqueous solutions the primary pathway for oxidation of organic compounds begins with the oxidation of water or OH^- ions that proceeds according to (5):

$$MO_x + H_2O \rightarrow MO_x(^\bullet OH) + H^+ + e^- \qquad (1)$$

$$MO_x + OH^- \rightarrow MO_x(^\bullet OH) + e^- \qquad (2)$$

where MO_x is the reactive site on the anode and $^\bullet OH$ is a hydroxyl radical that is physically adsorbed on the electrode surface. On active electrodes the physically adsorbed $^\bullet OH$ may react with the active site to produce a higher oxide, MO_{x+1}, according to (6):

$$MO_x(OH^\bullet) \rightarrow MO_{x+1} + H^+ + e^- \quad (3)$$

Both physically adsorbed $^\bullet OH$ and the MO_{x+1} are considered active oxygen species capable of oxidizing organic compounds. The active oxygen species may not only react with the desired organic compound, but may also decompose to dioxygen. On nonactive electrodes oxygen evolution occurs via the decomposition of adsorbed $^\bullet OH$ to form O_2 according to (5):

$$MO_x(^\bullet OH) \rightarrow 0.5\, O_2 + H^+ + e^- + MO_x \quad (4)$$

On active electrodes the higher oxide simply decomposes to regenerate the reactive site according to:

$$MO_{x+1} \rightarrow 0.5\, O_2 + MO_x \quad (5)$$

The nature of the electrode material and the electrode potential determine the activity of the electrode for organic compound oxidation. Electrodes coated with PbO_2, SnO_2 or other oxidation catalysts not capable of forming higher oxides are considered inactive electrodes. Anodes coated with IrO_2, RuO_2 or other catalysts capable of forming higher oxides are considered active electrodes.

Organic compound oxidation by active electrodes generally proceeds according to (6):

$$R + MO_{x+1} \rightarrow RO + MO_x \quad (6)$$

where R is the organic species. These oxidation reactions are usually selective and are not capable of completely mineralizing the organic compounds to CO_2. In contrast, oxidations by nonactive electrodes are capable of complete mineralization to carbon dioxide. This difference is due primarily to the fact that hydroxyl radicals are stronger oxidizing agents than MO_{x+1} species.

Organic compound oxidation by hydroxyl radicals usually begins via the abstraction of a hydrogen atom to form an organic radical according to:

$$RH + {}^{\bullet}OH \rightarrow R^{\bullet} + H_2O \qquad (7)$$

Once formed, these organic radicals can react with O_2 produced at the anode or already present in the water, according to:

$$R^{\bullet} + O_2 \rightarrow ROO^{\bullet} \qquad (8)$$

These reactions with dioxygen are desirable since they take advantage of the electrical energy already expended in electrochemically generating the O_2. In addition to further attack by ${}^{\bullet}OH$, these organic radicals may abstract a hydrogen atom to form an organic hydroperoxide according to:

$$ROO^{\bullet} + R'H \rightarrow ROOH + R''^{\bullet} \qquad (9)$$

These hydroperoxides are unstable and often decompose to organic compounds with fewer carbon atoms (5).

Electrochemical oxidation of organic compounds requires an anode that is stable under anodic polarization, and that has a low catalytic efficiency for oxygen evolution. Noble metals, such as platinum, are resistant to oxidation, but they have high catalytic efficiencies for oxygen evolution and are prone to fouling (6, 7). Dimensionally stable anodes, such as titanium coated with active or inactive catalysts, are less prone to oxygen evolution and fouling, but they do suffer from leaching of the catalyst from the electrode surface.

Boron Doped Diamond (BDD) Electrodes

The development of inexpensive chemical vapor deposition techniques for growing synthetic diamonds has resulted in widespread interest in polycrystalline diamond films for industrial applications (8). The films are often prepared on p-silicon substrates that have been polished with a diamond containing paste, and thereby contain adsorbed diamond crystals to serve as nucleation sites. Microwave radiation or a hot filament is used to decompose a low pressure gas mixture of methane and hydrogen. Upon decomposition, the methyl and atomic hydrogen radicals chemisorb at the nucleation sites and propagate growth of the film. Boron doping is often accomplished by adding trace amounts of B_2H_6 to the seed gas mixture. The boron atoms substitute for carbon in the diamond lattice and serve to increase the electrical conductivity of the diamond film. The resulting films are polycrystalline and may have resistivities as low as $< 0.1 \, \Omega$-cm (9).

The advantages of BDD electrodes over other materials includes their: 1) high resistance to fouling by chemisorbed metals or other impurities; 2) very low catalytic activity for oxygen evolution; 3) high mechanical strength and resistance to chemical attack; 4) high dimensional stability under anodic polarization; and 5) hydrophobicity. Hydrophobic electrodes are desirable for reactions of organic compounds in aqueous systems because of increased organic compound adsorption.

Materials and Methods

Batch experiments were conducted in a stirred, 20 mL, sealed, glass cell containing either ethanol or aqueous solutions. Sodium chloride was added to each solvent to provide electrical conductivity, and sodium hydroxide was added to the aqueous solutions in order to increase the solubility of triclosan. The pK_a value for triclosan is 7.9 (*10*), and all experiments in aqueous solutions were conducted at a constant pH value of 12. The working electrode was a boron doped diamond (BDD) film on a silicon substrate (CSEM, Neuchâtel, Switzerland) with a nominal surface area of 1 cm^2. A stainless steel wire encased in a Nafion® (DuPont) sheath was used as the counter electrode, and the reference electrode was Hg/Hg$_2$SO$_4$ (EG&G, Oak Ridge, TN).

Experiments were also conducted in a DiaCell® (CSEM) flow-through reactor. The anode consisted of a 100 mm diameter BDD film on p-silicon (CSEM), the cathode was a 100 mm diameter zirconium disk (Aesar, Ward Hill, MA), and the reference electrode was a Ag/AgCl microelectrode (EG&G). Aqueous or ethanol solutions were pumped through the reactor using a liquid chromatography pump at flow rates ranging from 0.63 to 10 mL/min. These flow rates resulted in mean hydraulic detention times in the cell ranging from 123 to 8 minutes.

In all experiments the voltage or current was controlled using a model 273A potentiostat (EG&G) and M270 software. Each experiment was repeated two or three times, and good reproducibility was observed. All potentials are reported with respect to the standard hydrogen electrode (SHE). Triplicate analyses for triclosan and its oxidation products were performed using liquid chromatography and gas chromatography-mass spectrometry.

The Microtox® assay was used to monitor the decline in toxicity of the triclosan solutions as a function of the electrolysis time. This test monitors the decline in respiration of the luminescent bacterium, *vibrio fischeri*. The procedures and equipment used in the Microtox® assays have been published elsewhere (*11*). Relative toxicity values for the electrolyzed solutions were calculated by normalizing the EC_{50} for each sample by the EC_{50} of the starting triclosan solution.

Results and Discussion

Batch Reactor

Figure 2 shows cyclic voltammetry scans in a 4 mM aqueous solution of triclosan at a pH value of 12. The first scan clearly shows an oxidation peak for triclosan centered at a potential of 1.1 V. The absence of any cathodic peak on the reverse scan shows that triclosan oxidation was irreversible. Subsequent scans showed oxidation peaks of diminishing magnitude, and by the fourth scan no oxidation peak was discernable. The diminishing peak amplitude with each scan suggests that triclosan oxidation produced a passivating film on the electrode surface. Passive films resulting from free radical polymerization reactions have been found to form during oxidation of phenolic compounds (*12*).

The passivating film could be removed by polarizing the electrode to higher potentials. Figure 3 shows the first scan for passivated BDD electrodes that were conditioned at current densities of 5 and 15 mA/cm^2. Conditioning at anodic current densities of 5 and 15 mA/cm^2 for 100 seconds was able to recover only a fraction of the initial electrode activity. However, conditioning for 1000

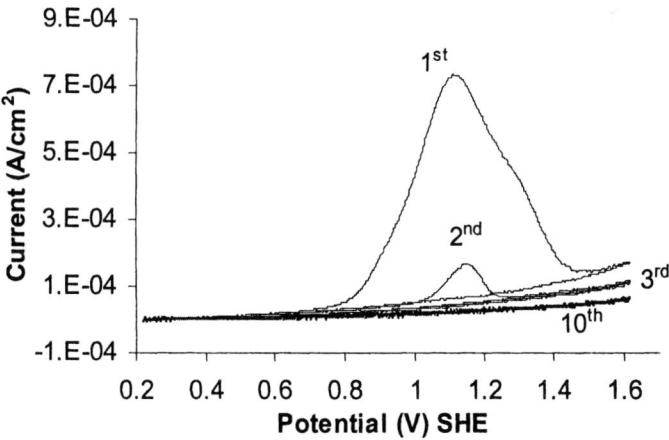

Figure 2. Cyclic voltammetry scans with BDD electrode in 4 mM triclosan solution at a pH value of 12.

seconds at an anodic current density of 15 mA/cm² was able to restore the electrode activity to its initial state. The cleaning effect can be attributed to oxidation of the passivating film by ˙OH generated from water oxidation.

No oxidation peaks for triclosan were observed in cyclic voltammetry scans in ethanol solutions. At all potentials the anodic currents in the presence of triclosan were lower than those in the blank ethanol solutions. This can be attributed to partial inactivation of the surface by oxidation product films. In extended electrolysis experiments conducted at a current density of 5 mA/cm², nearly complete inactivation of the electrode for triclosan destruction was observed by 2 hours elapsed, as shown in Figure 4. However, continuation of the electrolysis at a current density of 15 mA/cm² was able to partially restore the triclosan destruction activity.

Figure 5 shows triclosan concentrations as a function of electrolysis time in ethanol solutions at current densities of 15 and 25 mA/cm². The linear decline in triclosan concentrations with time indicates that the destruction kinetics were zeroth order in triclosan concentration. This can be explained by a reaction that was limited by the oxidation current. This conclusion is supported by comparison of the removal rates at 15 and 25 mA/cm². The zeroth order rate constant at 25 mA/cm² of 2.2×10^{-6} mol/min is a factor of 1.47 greater than the rate constant of 1.5×10^{-6} mol/min at 15 mA/cm². This ratio is close to the factor of 1.67 difference between the two current densities.

In contrast to the zeroth order removal kinetics in ethanol, more complex kinetic behavior was observed for triclosan electrolysis in aqueous solutions. Figure 6 compares triclosan removal rates at current densities of 5 and 15

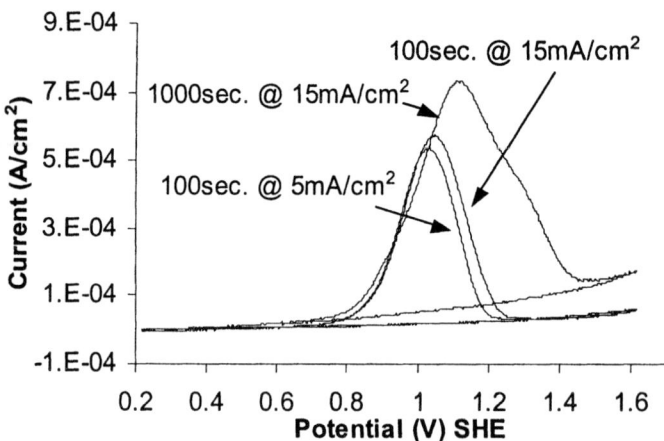

Figure 3. The effect of anodic conditioning on recovery of the voltammetric response for triclosan oxidation at a pH value of 12.

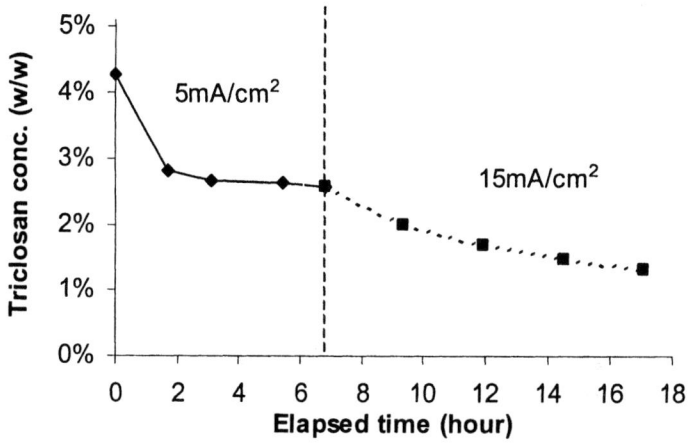

Figure 4. Galvanostatic electrolysis of triclosan in ethanol solutions at anodic current densities of 5 and 15 mA/cm^2.

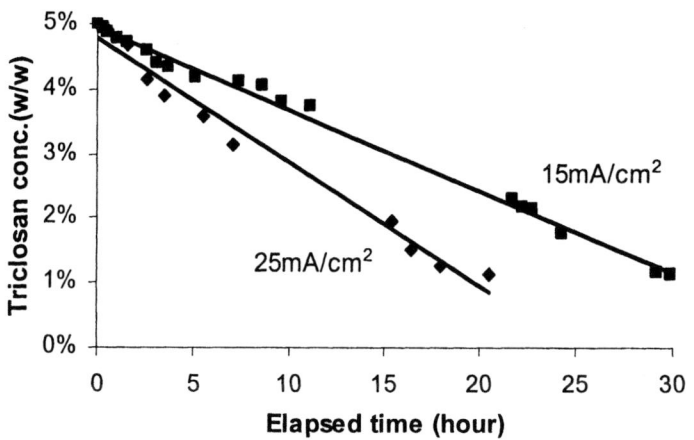

Figure 5. Triclosan concentrations in ethanol solutions as a function of the electrolysis time at current densities of 15 and 25 mA/cm^2.

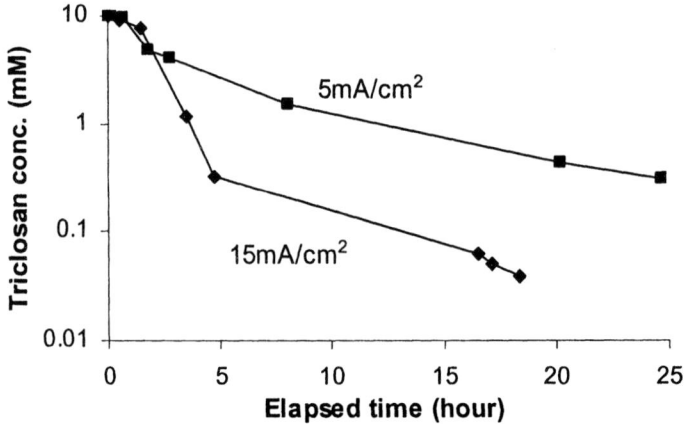

Figure 6. Triclosan concentrations in aqueous solutions at a pH value of 12 for current densities of 5 and 15 mA/cm².

mA/cm². Over the first two hours elapsed the removal rates were similar at both current densities. This can likely be attributed to reaction rates that were limited by diffusive mass transport to the electrode surface. Between 2 and 5 hours elapsed the triclosan removal rates were faster than those before 2 hours elapsed. This can likely be attributed to a free-radical chain reaction mechanism involving organic radicals formed by triclosan oxidation and triclosan itself. Attack of triclosan by organic free-radical species in solution may be expected to increase the triclosan removal rate because diffusion of triclosan to the electrode surface for reaction with adsorbed ˙OH is not required. The declining removal rates for triclosan concentrations below 0.5 mM may arise from competitive oxidation of triclosan byproducts. Reaction of partial degradation products with ˙OH adsorbed on the electrode surface must have decreased the availability of ˙OH for reacting with triclosan.

Flow-Through Reactor

Effluent triclosan concentrations from the flow-through reactor operated at a current density of 6 mA/cm² with ethanol and aqueous solutions are shown in Figure 7. Much faster triclosan degradation kinetics were observed in the aqueous solutions than in ethanol. This can be explained by the fact that solvent oxidation in the aqueous solution produced the strong oxidant ˙OH, while ethanol oxidation produced only a weak oxidant. Oxidation of ethanol occurs via abstraction of a hydrogen atom according to (*13*):

$$CH_3CH_2OH \rightarrow CH_3CH_2O^\bullet + H^+ + e^- \tag{10}$$

Although the resulting ethoxy radicals could react with triclosan, they are much less reactive than hydroxyl radicals. The recombination of ethoxy radicals may also account for slower triclosan destruction rates in the ethanol solutions.

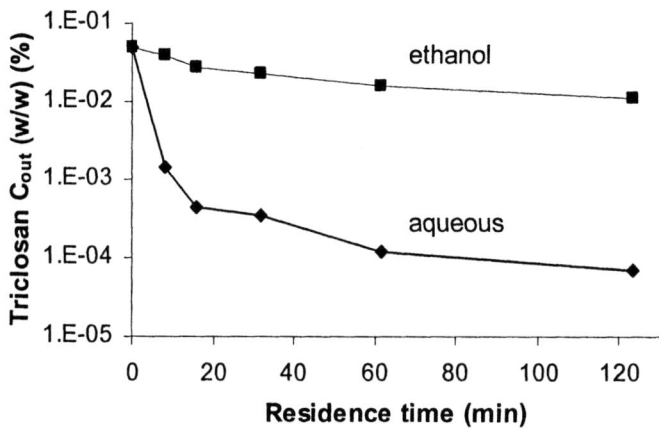

Figure 7. Effluent triclosan concentrations from the flow-through reactor in ethanol and aqueous solutions at a current density of 6 mA/cm^2.

The reaction products in ethanol confirm that ethoxy radical recombination was a major pathway for terminating free-radical chain reactions. Figure 8 shows that the two most preponderant reaction products in the ethanol solutions resulted from recombination of ethoxy radicals, and from reactions of ethoxy radicals with ethanol. The other major products in the ethanol solutions suggest that breaking the ether bond between phenyl groups was easier than opening the aromatic rings. The products produced in the aqueous electrolyses also showed that breaking the ether linkage was facile. Substituted acetic acids were the two most common products in the aqueous solutions. This is consistent with previously reported results from phenol oxidation in which a variety of one to three carbon aliphatic acids are produced (*14*). The preponderance of acetic acid is consistent with prior reports that acetic acid is the most recalcitrant of these acids to further oxidation (*6*).

All reaction products in both ethanol and aqueous solutions were much less toxic than triclosan itself. Microtox® EC$_{50}$ values were calculated for effluent samples from the flow-through reactor. Figure 9 shows how the decline in toxicity of the solutions closely parallels the decline in triclosan concentration.

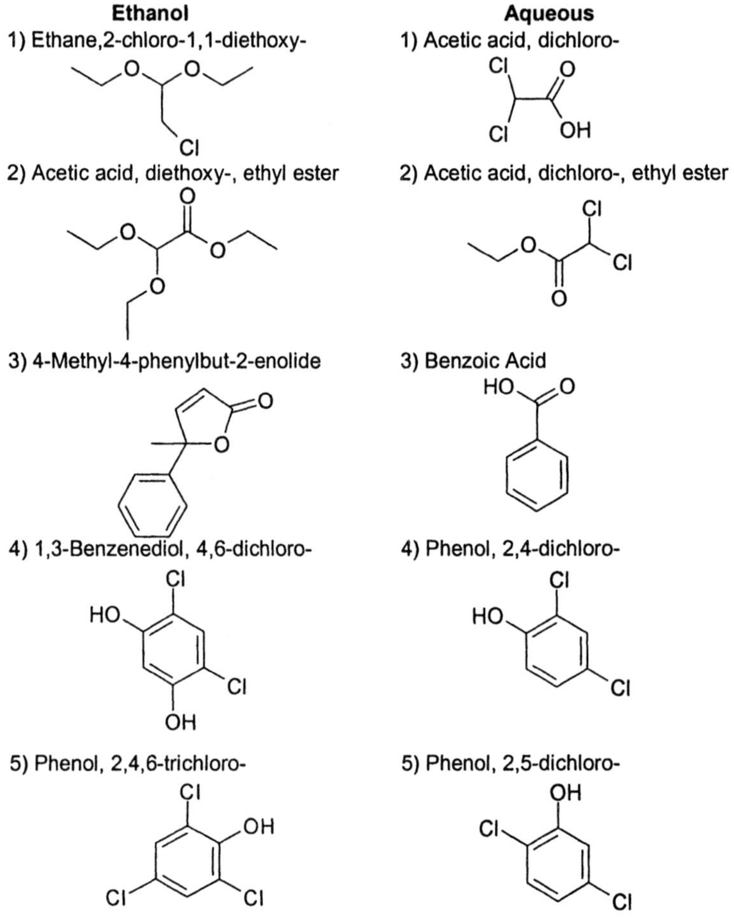

Ethanol
1) Ethane,2-chloro-1,1-diethoxy-

2) Acetic acid, diethoxy-, ethyl ester

3) 4-Methyl-4-phenylbut-2-enolide

4) 1,3-Benzenediol, 4,6-dichloro-

5) Phenol, 2,4,6-trichloro-

Aqueous
1) Acetic acid, dichloro-

2) Acetic acid, dichloro-, ethyl ester

3) Benzoic Acid

4) Phenol, 2,4-dichloro-

5) Phenol, 2,5-dichloro-

Figure 8. Major byproducts of triclosan electrolysis in ethanol and aqueous solutions after 2 hours of electrolysis at a current density of $6mA/cm^2$.

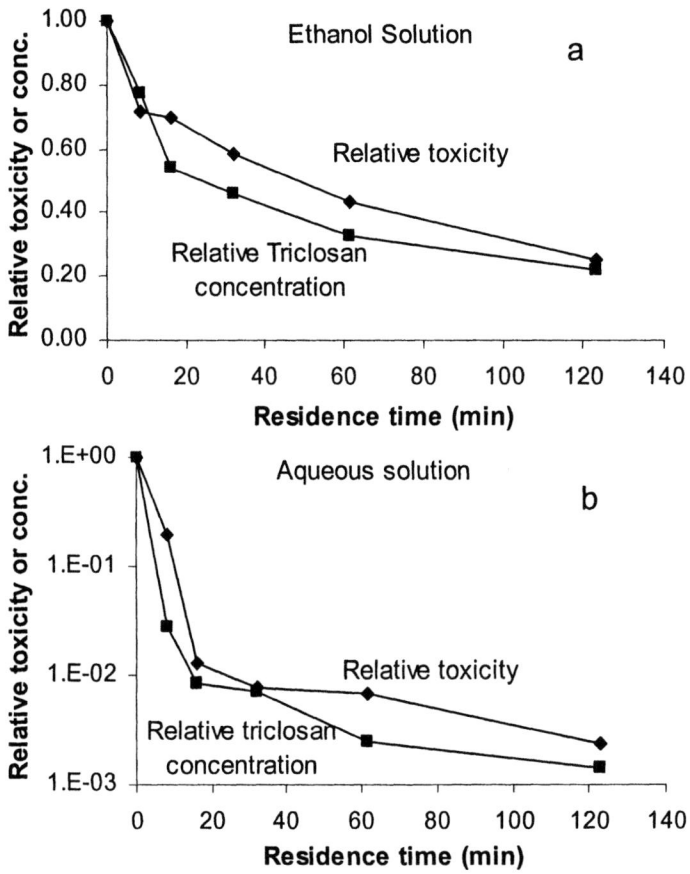

Figure 9. Normalized toxicity and triclosan concentrations in ethanol and aqueous solutions during electrolysis at a current density of 6 mA/cm^2.

This suggests that residual triclosan concentrations will be the major factor in determining the biodegradability of the partial degradation products.

Conclusion

Triclosan removal rates in ethanol solutions were zeroth order with respect to the triclosan concentration, and were proportional to the current density. This indicates that triclosan removal rates were limited by the cell current. In aqueous solutions the triclosan removal kinetics were more complex, and were more than two orders of magnitude faster than in the ethanol solutions. A two order of magnitude reduction in triclosan concentration and relative toxicity could be achieved in less than 20 minutes in the flow-through reactor. This suggests that aqueous electrolysis may be a feasible process for destroying aging and contaminated stockpiles, or for treating solutions generated during triclosan manufacture and purification.

References

1. McMurry, L. M.; Oethinger, M.; Levy, St. B. *Nature* **1998**, *394*, 531-532.
2. Lindstrom, A.; et al. *Environ. Sci. Technol.* **2002**, *36*, 2322-2329.
3. Kolpin, D. W.; et al. *Environ. Sci. Technol.* **2002**, *36*, 1202.
4. Adolfsson-Erici, M.; Pettersson, M.; Parkkonen, J.; Sturve, J. *Chemosphere* **2002**, *46*, 1485-1489.
5. Comninellis, Ch. *Electrochimica Acta* **1994**, *39*, 11/12, 1857-1862.
6. Gyorgy, F.; Gandini, D.; Comninellis, C.; Perret, A.; Haenni, W. *Electrochemical and Solid State Letters* **1999**, *2*, (5), 228-230.
7. Comninellis, Ch.; Nerini, A. *J. Appl. Electrochem.* **1995**, *25*, 23
8. Dischler, B.; Wild, C. (Eds.) *Low Pressure Synthetic Diamond Manufacturing and Applications, Springer Series in Materials Processing*, Springer-Verlag: Berlin, 1998.
9. Xu, J.; et al. *Analytical Chem.* **1997**, *69*, 591A-597.
10. Pemberton, R. M.; Hart, J. P. *Analytica Chimica Acta* **1999**, *390*, 107-115.
11. J.M. Ribo; Kaiser K.L.E. *Tox. Assess.* **1987**, *2*, 305-323
12. Iniesta, J.; Michaud, P. A.; Panizza, M.; Cerisola, G.; Aldaz, A.; Comninellis, Ch. *Electrochimica Acta* **2001**, *46*, 3573.
13. Kim, J.W.; Park, S.M. *J. Electrochem. Soc.* **1999**, *146*, 3, 1075-1080.
14. Wu, Z.; Zhou, M. *Environ. Sci. Technol.* **2001**, *35*, 2698-2703.

Chapter 9

Detoxification of Pesticide in Water Using Solar Photocatalysis

S. Malato[1] and A. Agüera[2]

[1]Plataforma Solar de Almería-CIEMAT, Ctra. Senés Km. 4, Carretera Senes, Room 4, Tabernas, 04200 Spain
[2]Pesticide Residue Research Group, Department of Analytical Chemistry, University of Almería, 04071 Spain

This paper deals with the use of sunlight to produce •OH radicals by photocatalysis. The systems necessary for performing pilot-plant-scale experiments based on compound parabolic collectors (CPCs) are described. It outlines the basic components of these plants (reflective surface and glass absorber). It has been demonstrated the total disappearance and mineralisation of pure and commercial pesticides dissolved in water (C_0 = 50 mg/L) using 0.2 g/L of TiO_2. Besides, it underlines the importance of: (i) using acute toxicity bioassays, for stating biocompatibility of the treated water with the environment and (ii) using photocatalysis as a pre-treatment step, if the intermediates resulting from the reaction are readily degraded by microorganisms (biotreatment).

Introduction

The use of pesticides has risen dramatically, with the production nearly doubled every five years since 1975. UN reports estimates that of all pesticides used in agriculture, less than 1% actually reaches the crops. The agriculture is increasingly and widely using pesticides, with 400.000 Tm used in the European Union just in 2000. This results in the uncontrolled disposal of used products that will produce contaminated soils and waters close to the contaminant source. Their persistence in natural waters (1) has led to a search for a method to degrade them into environmentally compatible compounds. Unlike the low-level contamination involved in drinking water, wastewater from agricultural or industrial activities may be highly contaminated. The major sources of pollution are wastewater from agricultural industries, pesticides formulating and manufacturing plants. Wastewater from those sources may contain pesticides at levels as high as several hundred of mg/L. The main characteristics of them are its extreme toxicity, low volume and well-defined location. Such sources may be ideally treated in small-scale treatment units. As consequence, low cost and at hand technologies are strongly urged to be developed to on site treatment.

A wide range of pesticides can be classified as Persistent Organic Compounds (2), therefore they resist photolytic, biological and chemical degradation. In such cases, it is necessary to adopt much more effective systems (3): air stripping, adsorption on granulated activated carbon, incineration, ozone, oxidation, etc. Air stripping and adsorption, merely transferring toxic materials from one medium to another, is not a long-term solution (4). Incineration (5) is capable of converting toxics to carbon dioxide, water and inorganic acids, but negative public perception has very often prevented its implementation. Ozone is not able very often to completely destroy pesticides to harmless compounds (6). Much research has addressed chemical oxidation in the last decade (7-10) pointing out the prominent role of a special class of oxidation techniques defined as AOP (Advanced Oxidation Processes), which usually operate at or near ambient temperature and pressure (see Fig. 1).

AOPs, while making use of different reaction systems, are all characterized by the same chemical feature: production of OH radicals ($^{\bullet}$OH). These radicals are extraordinarily reactive (oxidation potential 2.8 V) and attack most organic molecules. They are characterized by their not very selective attack, which is a useful attribute for an oxidant used in pollution problems. Only wastes with small COD contents (a few g/L) can be suitably treated by these techniques since higher contents would require the consumption of too large amounts of expensive reactants. Methods based on UV, H_2O_2/UV, O_3/UV and H_2O_2/O_3/UV combinations use photolysis of H_2O_2 and ozone to produce the hydroxyl radicals (11). Besides, other methods like heterogeneous photocatalysis and homogeneous photo-Fenton are based on the use of a wide band gap semiconductor and in the addition of H_2O_2 to Fe^{2+} salts and by irradiation with UV-VIS light. Both processes are of special interest since sunlight can be used

- H_2O_2/Fe^{2+} (Fenton): $Fe^{2+} + H_2O_2 \rightarrow Fe^{3+} + OH^- + OH^{\bullet}$
- H_2O_2/Fe^{2+} (Fe^{3+})/UV (photo-Fenton): $Fe^{3+} \xrightarrow{h\nu} Fe^{2+} + HO^{\bullet}$ } CATALYSIS + SUN
- $TiO_2/h\nu/O_2$ (Photocatalysis): $TiO_2 \xrightarrow{h\nu} e^- + h^+$
  $h^+ + H_2O \rightarrow {}^{\bullet}OH + H^+$
- O_3/H_2O_2: $H_2O_2 \xleftrightarrow{H^+} HO^- + O_3 \rightarrow O_2 + HO_2^-$; $HO_2^- + O_3 \rightarrow HO_2^{\bullet} + O_3^{\bullet-}$
  $HO_2^{\bullet} \Leftrightarrow H^+ + O_2^{\bullet-}$; $O_2^{\bullet-} + O_3 \rightarrow O_2 + O_3^{\bullet-}$; $O_3^{\bullet-} + H^+ \rightarrow HO_3^{\bullet}$
  $HO_3^{\bullet} \rightarrow HO^{\bullet} + O_2$; $HO^{\bullet} + O_3 \rightarrow HO_2^{\bullet} + O_2$
- O_3/ UV: $O_3 \xrightarrow{h\nu} O^1(D) + O_2$; $O^1(D) + H_2O \rightarrow H_2O_2$; $H_2O_2 \xrightarrow{h\nu} 2HO^{\bullet}$
- H_2O_2/UV: $H_2O_2 \xrightarrow{h\nu} 2OH^{\bullet}$

Figure 1. Production of $^{\bullet}OH$ through AOPs.

for them (12). The main problem with the AOPs is their high cost (expensive reactants as H_2O_2 and generation of UV). So, the future applications of these processes could be improved through the use of catalysis and solar energy.

Detoxification By Solar Photocatalysis

Heterogeneous TiO_2 photocatalysis and homogeneous photo-Fenton, are the AOPs for which the solar technologies have been most extensively studied and developed (13). In the last few years, solar photocatalysis processes have shown to be an important alternative for pesticide degradation in the field of AOPs (14, 15) with extended research performed at pilot-plant scale (16, 17). The renewed interest of researchers (18) in the photo-assisted Fenton processes (classic old reactive system, discovered by Fenton (19) in the last century) is today underlined by a significant number of studies devoted to wastewater treatment (20). Production of OH radicals by Fenton reagent occurs by means of the addition of H_2O_2 to Fe^{2+} salts. The degradation rate of organic pollutants is strongly accelerated with photo-Fenton by irradiation with UV-VIS light. Under these conditions, the photolysis of Fe^{3+} complexes allows Fe^{2+} regeneration and the occurrence of Fenton reactions due to the presence of H_2O_2 and iron may be considered a true catalyst. Hydroxyl radicals can also be generated with a semiconductor that absorbs radiation when in contact with water and generates pairs of valence-band holes and conduction-band electrons. References and patents related to the heterogeneous photocatalytic removal of toxic compounds from water published during the last decade can be counted in the thousands (21). Whenever semiconductor materials have been tested under comparable conditions, TiO_2 has generally been demonstrated to be the most active (22).

Solar Detoxification Technology

Laboratory research on photocatalysis has mostly been performed with experimental devices in which efficiency was not as important as obtaining appropriate conditions that would permit exhaustive knowledge of all the important parameters. Usually, the UV-light is provided by a lamp introduced in a water-cooled envelope or by solar simulators. All the IR beams, which could heat the suspensions, are removed by a water cell. This is correct but not always sufficient to attempt a change of scale. Nevertheless, the design procedure for a pilot system requires the selection of a reactor, catalyst, reactor-field configuration (series or parallel), treatment-system mode (once-through or batch), flow rate, pH control, etc., so a pilot plant has to be as versatile as possible and provide sufficient confidence in the experiments carried out in it (17). In Fig. 2, a detailed drawing of a solar photocatalytic plant is given. Usually, a detoxification pilot plant is constructed with several solar collectors (see below). All the modules are connected in series, but with valves that permit to bypass any number of them. All the connection tubes and valves are strongly resistant to chemicals, weather-proof and opaque, in order to avoid any photochemical effect outside of the collectors. The most important sensors required for the system are temperature, pressure and dissolved oxygen (at least in the reactor outlet). An injection system of oxygen at the reactor inlet allows it to be added to the reactor. A UV-radiation sensor must be placed in a position where the solar UV light reaching the photoreactor can be measured. Solar detoxification pilot plants are frequently operated in a recirculating batch mode. In this situation, the fluid is continuously pumped between the reactor and a tank in which no reaction occurs, until the desired degradation is achieved.

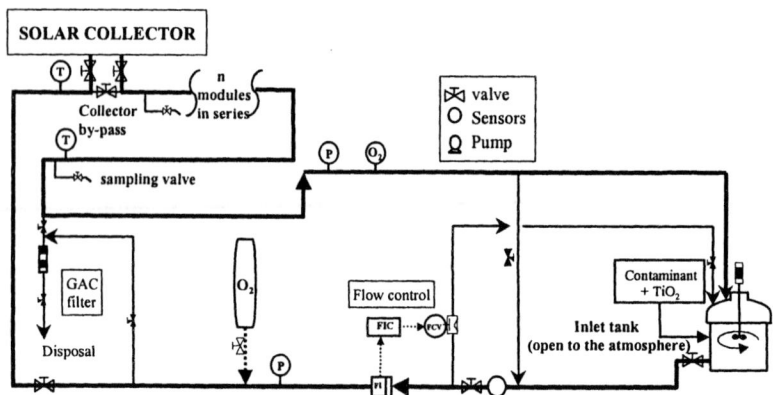

Figure 2. Photocatalytic Pilot Plant scheme (recirculating mode is shown).

Non-concentrating solar collectors are the choice for solar photocatalytic applications. They are more efficient than concentrator-based systems due to the use of both direct and diffuse UV light and their intrinsic simplicity (23). Compound parabolic collectors (CPCs) are static collectors with a reflective surface formed by two connected parabolic mirrors with an absorber tube in the focus (see Fig. 3a) and have been found to provide the best optics for low concentration systems (24, 25). It has no tracking system and the design permits the solar rays to be reflected onto the focus (absorber) and low concentration ratio is attained. Thanks to the reflector design, almost all the UV radiation arriving at the CPC aperture area can be collected and is available for the process in the reactor. The light reflected by the CPC is distributed around the back of the tubular photoreactor (see Fig. 3b). Due to the ratio of CPC aperture to tube diameter, the incident light is very similar to that of simple tubular photoreactor.

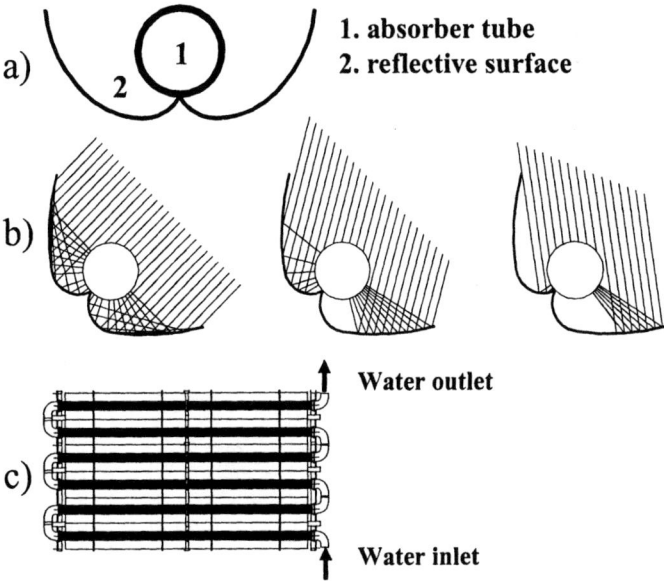

Figure 3. Schematic drawing of CPCs.

All the experiments shown in this paper were carried out under sunlight in CPCs at the Plataforma Solar de Almería (PSA, latitude 37°N, longitude 2.4°W). The pilot plants (17) are made up of different solar collectors, tanks and pumps. Each collector (see Fig. 3c) consists of Pyrex tubes (installed in the axis of the CPC) connected in series and mounted on a fixed platform tilted 37° (local latitude). The water flows directly from one module to another and finally into a tank. The total volume (V_T) of the reactor is separated in: total irradiated volume

(Pyrex tubes, V_i) and the dead reactor volume (tank + connecting tubes). Solar UV was measured by a global UV radiometer (KIPP&ZONEN, model CUV3), mounted on a platform tilted 37° (the same angle as the CPCs), which provides data in terms of incident W_{UV} m^{-2}. This gives an idea of the energy reaching any surface in the same position with regard to the sun. With Eq. 1, combination of the data from several days' experiments and their comparison with other photocatalytic experiments is possible.

$$t_{30W,n} = t_{30W,n-1} + \Delta t_n \frac{UV}{30} \frac{V_i}{V_T}; \quad \Delta t_n = t_n - t_{n-1} \qquad (1)$$

where t_n is the experimental time for each sample, UV is the solar ultraviolet radiation measured during Δt_n, and t_{30W} is a "normalized illumination time". In this case, time refers to a constant solar UV power of 30 W m^{-2} (typical solar UV power on a perfectly sunny day around noon). Three different plants based on CPCs are installed at PSA. The larger has 3 modules (collector surface 3 m^2 each) mounted on a fixed platform tilted 37°. Total plant volume is 247 L and photoreactor (illuminated) volume 108 L. Besides, other 2 small twin-systems (for performing comparative tests) are available, each having three collectors (1.03 m^2 each), one tank and one pump. Each collector consists of eight Pyrex tubes. Total volume of the reactor is 40 L and total irradiated volume is 22 liters.

Solar Detoxification of Pesticides

Up to now, practical applications of solar technologies have been studied and developed most intensively for TiO_2 photocatalysis and photo-Fenton. Treatment of industrial wastewater, in spite of inefficient production of hydroxyl radicals and slow kinetics (26), seems to be one of the most promising fields of application of solar detoxification. The only really general rule is that there is no general rule at all, each real case being completely different. Consequently, preliminary research is always required to optimize the best option for any specific problem, on a nearly case-by-case basis. In this section, an attempt will be made to summarize the results obtained with different commercial and pure pesticides (27-41). They have been selected because they have different structures representative of a wide range of modern pesticides (containing different heteroatoms) and they are easily soluble in water (of special interest because of their extremely easy transport in the environment,). It is necessary to remark that this section has been devoted only to TiO_2 because the limited extent of this contribution, but dozens of articles have been published about photo-Fenton treatment of pesticides, several of them listed in the references list

(8,9,12,14,18-20,31,36,38-41). Concerning photo-Fenton, the general conclusion is that the same installation as for TiO_2 can be used, and its efficiency and applicability must be considered also in a case-by-case basis.

Pure Pesticides

The OH radicals are able to initiate a series of oxidative reactions that can give rise to the complete mineralisation of the pesticides. But, obviously the performance of these treatments depend on the nature of the compounds as well as on the selected operational parameters. In Fig. 4 it is shown the photocatalytic treatment of five different pesticides: imidacloprid, methomyl, diuron, formetanate and pyrimethanil. As observed in Fig. 4, all pesticides can be successfully destroyed by photocatalysis. However, in AOPs transformation of the parent organic compound is desirable but the principal objective is to mineralize all pollutants. The effectiveness of degradation is not demonstrated only because the entire initial compound is decomposed. In all cases, many new organic compounds have appeared because TOC remains high after parent compound total disappearance and total mineralisation (i.e., complete disappearance of TOC complete and complete release of heteroatoms as inorganic acids) can be attained only after very long irradiation.

Photocatalytic disappearance with TiO_2 follows apparent first-order kinetics when initial concentration is low enough. It should be emphasized that all pesticides decompose giving rise to by-products, which could also be competitive on the surface of the TiO_2. Their concentration varies throughout the reaction up to their mineralisation and thus, the following equation (based on the Langmuir-Hinselwood kinetic model and commonly used for describing photocatalysis by TiO_2) could describe the kinetics:

$$r = \frac{k_r KC}{1 + KC + \sum_{i=1}^{n} K_i C_i (i=1, n)} \quad (2)$$

where k_r is the reaction rate constant, K is the reactant adsorption constant, C is concentration at any time, K_i is the by-products adsorption constant and C_i the by-products concentration at any time. With this model, kinetics of the results shown in Fig. 4 (left) could be known, but, as mineralisation (Fig.4, right) does not follow simple models, overall reaction rate constants cannot be calculated. The complexity of the results, of course, is caused by the fact that the TOC is a sum parameter often including several hundred products that undergo manifold reactions. Looking at these results and considering the disparity of results obtained, it is possible to envisage that experimentation with pure compounds

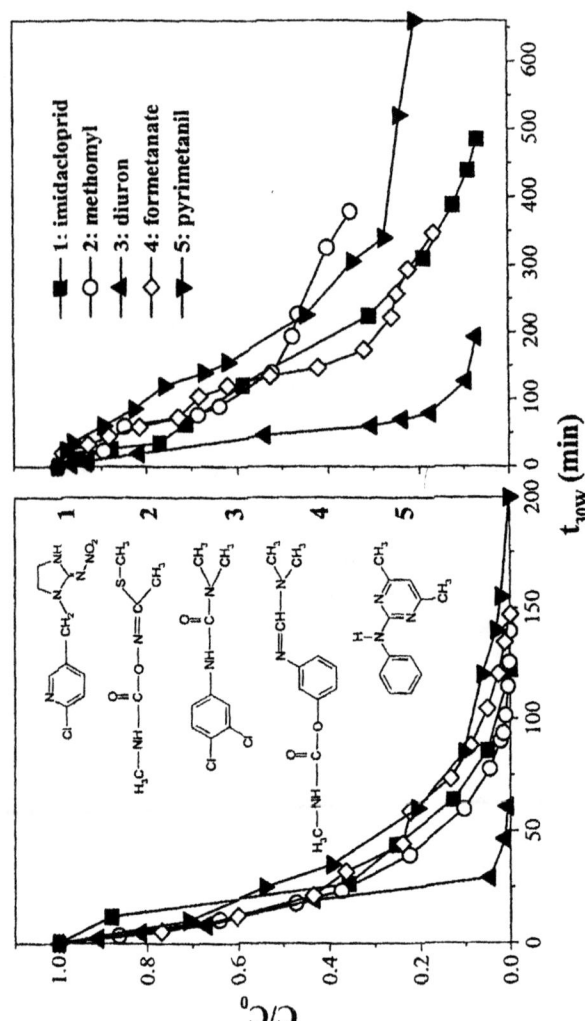

Figure 4. *Disappearance (left) and mineralisation (right) of different pesticides at $C_0 = 50$ mg/L using TiO_2 (200 mg/L) in the pilot plant shown in Fig. 2.*

could be useful for clarifying different questions (degradation pathways, main degradation products, etc.) but not for designing a suitable treatment for a real wastewater. In these cases the logical choice is to test commercial pesticides.

A good example it is observed comparing Fig. 4 and Fig. 5 for imidacloprid (C_0=50 mg/L) and its commercial formulation (Confidor®). A rapid initial decrease is observed between 0 and 100 min in the case of Confidor solution. The half-life time was about 1 h and the complete disappearance took place after 270 min. When technical imidacloprid was used, a faster degradation rate was observed achieving total disappearance of imidacloprid after only 125 minutes with a half-life time of about 25 min. This fact is a consequence of the competence existing between the pesticide and the formulation components for the reaction with hydroxyl radicals over TiO_2 surface. Initial TOC higher than 120 mg/L in the Confidor solution confirmed the presence of additional ingredients corresponding to the commercial formulation. These compounds produce a high decrease in the mineralisation rate. 50% abatement of the initial concentration took place after 500 min. Initial TOC concentration in the case of technical grade was 20 mg/L and 175 min were necessary to achieve similar results. An example of no different behavior between technical and commercial product has been presented in a recent work (35), where pyrimetanil photocatalytic degradation is compared with its formulation Scala®. In this case the commercial formulation does not contain other organic compounds.

Commercial pesticides

All tests shown in Fig. 5 have been carried out using commercial products, because the treatment must destroy not only active matter, but any other organic contained in the formulation as well. Unfortunately, very little information on the effects of adjutants in photocatalysis degradation is available because pesticides commercial formulations are usually patent protected. Fig. 5 summarizes tests carried out with 4 commercial pesticides. The initial TOC obtained from adding 100 mL of Vydate® (24% w/v oxamyl) to 247 L of water produced a TOC of almost 90 mg L^{-1}, of which only 37 mg L^{-1} come from oxamyl, and the rest come from the formulation. Likewise, with 62 mL of Confidor® (20% w/v imidacloprid) the initial TOC obtained is 130 mg L^{-1}, of which 21 mg L^{-1} of TOC are imidacloprid; adding 82 mL of Rufast® (15% w/v of acrinathrin), TOC was almost 50 mg/L, of which only 29 mg/L are from acrinathrin; a 0.33 mL/L Tamaron 50® (methamidophos 50% w/v)/water solution has a TOC of 100 mg/L (only 28 mg/L are from methamidophos).

Active matter may be observed to disappear very rapidly in comparison with the TOC. The disappearance of total organic carbon of all pesticides shows an induction period (no TOC degradation). During this period, formulation organics and intermediates produce more oxidized substances, but these are not mineralized. Mineralisation only occurs when the last step yields CO_2. This induction period differs between the different pesticides and it is difficult to

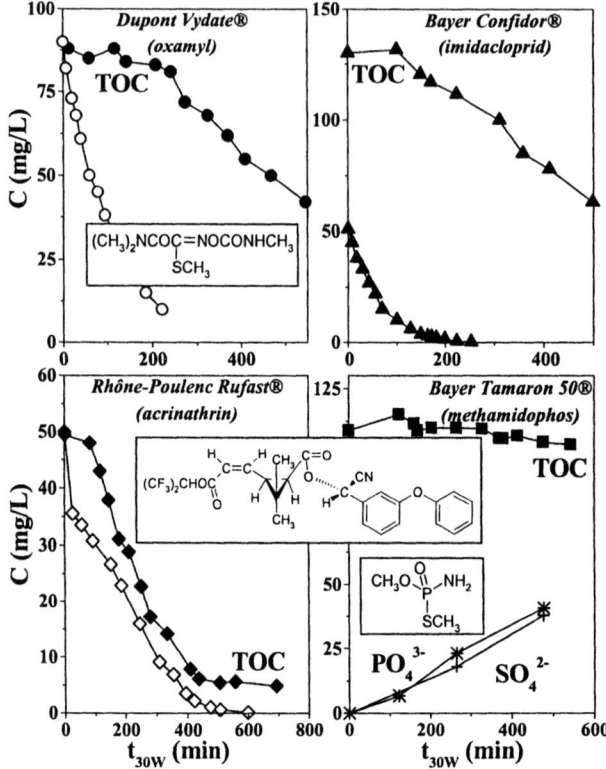

Fig. 5. Disappearance and mineralisation of different commercial pesticides as a function of t_{30w} (illumination time) using TiO_2 (200 mg/L)

predict. Tamaron 50 ® (methamidophos) is an example of no significant mineralisation: 10% initial TOC is degraded in 550 min irradiation. This is not surprising, because the initial TOC concentration is high and the •OHs are consumed during partial oxidation of all the organics present in the wastewater (methamidophos + formulation components). Nevertheless, PO_4^{3-} and SO_4^{2-} are produced throughout the experiments (almost at the same rate). It may be inferred that no other P or S compounds are present in the formulation because methamidophos generates the same quantity of PO_4^{3-} and SO_4^{2-}.

With the general objective of designing a treatment plant for wastewaters containing commercial pesticides, it should be remarked that the reaction rate is directly related to the size of the solar field, which is the most expensive component of a solar plant (42). For calculating the plant it is necessary to relate the mineralization rate with the solar photon flux reaching the collector surface. In the case of Vydate (oxamyl) the volume of water was 247 L and the collector used area 8.9 m². To degrade 90% (81 mg/L) of the initial TOC (90 mg/L) the illumination time has been t_{30W} = 910 min (it has been calculated considering the induction period, after it zero order kinetics and extrapolating data until 90%

conversion), therefore, the treatment capacity is: (81 mg/L) x (247 L) x (8.9 m^2)$^{-1}$ x (910 min)$^{-1}$ = 2.2 mg min^{-1} m^{-2}. The treatment capacity, calculated with an average intensity of 30 W$_{UV}$ m^{-2}, will be very useful for extrapolating the results to other different conditions. For example, the average yearly UV radiation at the Plataforma Solar de Almería (43) is 18.6 W$_{UV}$ m^{-2} (sunny hours 4380 h year^{-1}). The average yearly treatment capacity in this case would be approximately: (2.2 mg/min m^2) x (18.6/30) x (4380 h/year) x 60 x 10^{-6} = 0.35 kg m^{-2} year^{-1}. So, considering 1000 m^3/year and the same initial concentration, the reactor (similar to that shown in Fig. 3) size required is 231 m^2.

Photocatalysis of Pesticides: Outlook for the Future.

The proposed technology could also be applicable to other organic hazardous contaminants, such as solvents, detergents and a variety of industrial chemicals, which are capable of deep penetration into the soil and reach groundwater. Nevertheless, the process efficiency can be considered linearly dependent on the energy flux but only 5% of the whole solar spectrum is available for TiO$_2$ band-gap. A realistic assumption of solar collector efficiency of 75% and 1% for the catalyst (44) means 0.04% original solar photons are efficiently used in the process. From the standpoint of solar collecting technology, this is a rather inefficient process even considering for a high added-value application. Solar AOPs have the advantage over other AOPs of using sunlight and having as its main characteristic that it is an environmentally friendly technology. TiO$_2$ is a cheap photostable catalyst, and the process may run at ambient temperature and pressure. Additionally the oxidant, molecular oxygen (O$_2$) is the mildest. Therefore, in principle, the process involves a mild catalyst working under mild conditions with mild oxidants. However, as concentration and number of contaminants increase (as in complex mixtures of pesticides), the process becomes more complicated and challenging problems such as slow kinetics, low photoefficiency and unpredictable mechanisms need to be solved. It is clear that naked TiO$_2$ needs extra help to undertake practical applications and this may cause it to lose some of the charm of its mild operation.

Two basic lines of R&D (increasing quantum yield) have been working on modifying catalyst structure and composition and by adding electron acceptors (45). A third approach has focused on finding new catalysts able to work with band-gaps which better overlap the solar spectrum. There have been many attempts within the first and third approach, such as improving specific surface, doping and deposition with metal ions and oxides (21). Successful innovative catalyst compositions have been developed, but they have not been used in large-size plants because no "cheap" solution has yet been developed. Our experience in testing at large solar facilities and with different contaminants qualifies the use of electron acceptors as the most versatile way of improving reaction rates (for the moment), opening the opportunity to extend the using of heterogeneous photocatalysis to complicated waste water (46).

Contaminant treatment, in its strictest meaning, is the complete mineralisation (TOC = 0) of the contaminants, but, photocatalytic processes only make sense for hazardous non-biodegradable pollutants. When feasible, biological treatment is the cheapest treatment and also the most compatible with the environment. Therefore, biologically recalcitrant compounds could be treated with photocatalytic technologies until biodegradability is achieved, later transferring the water to a conventional biological plant. Such a combination reduces treatment time and optimizes the overall economics, since the solar detoxification system can be significantly smaller. Due to the kinetic mechanism (see Eq. 2), the first part of the photocatalytic process is the quickest. As it can be seen in the figures presented, the active component disappears after several minutes of irradiation, but TOC remains after hours of irradiation. Therefore, the use of AOPs as a pre-treatment step can be justified if the intermediates resulting from the reaction (more oxidized compounds as carboxylic acids, alcohols, etc) are readily degraded by microorganisms (47, 48). The feasibility of such a photocatalytic-biological process combination must always be assessed, because it could provide an important cost reduction by reducing the size of the necessary solar collector field. It must be taken into account that, as with most solar systems, economics of the water detoxification systems are dominated by their capital cost.

Determining the toxicity of the water, at different stages of AOP treatment, using different microorganisms is another way to decrease AOP operating costs. In this case, biocompatibility with the environment can be stated. Toxicity testing of the photocatalytically treated wastewater is therefore necessary, particularly when incomplete degradation is planned. Recently, the use of acute toxicity bioassays (49) has meant an important improvement in the evaluation of AOPs because of their reproducibility, adequate format for quick analysis, short analysis time, as well as well-defined analytical protocols. Assessment of the contaminant's effect involves summarizing data on the effects of the chemical on representative organisms and using these data to predict a no-effect concentration on a specific niche. Toxicity of a chemical is usually expressed as the effective concentration of the material that would produce a specific effect in 50% of a large population of test species (EC_{50}). The 'no observed effect concentration' (NOEC) is the concentration immediately below the lowest level eliciting any type of toxicological response in the study. All these analytical performance facts make the use of these acute bioassays very attractive for AOPs evaluation. Numerous bioassay procedures are now available (50, 51), however, if we consider that toxicity is a biological response, the values obtained by a single toxicity assay can be an insufficient measure of the adverse biological impact. Consequently, a battery of assays is recommended to be applied to assess toxicity adequately, and careful selection, of with regard to complimentarily is essential. A great confidence in the detoxification assessment is achieved, when two or more different bioassays representatives of different taxonomic groups point in the same direction (41).

References

1. Hayo, M.G. *Agric. Ecosys. Environ.* **1996**, *60*, 81.
2. Ritter, L.; Solomon, K.R.; Forget, J.; Stemeroff, M.; O'Leary, C. *Persistent Organic Pollutants*; Int. Prog. on Chem. Safety; UN Environ. Prog., 1996.
3. Zinkus, G.A.; Byers, W.D.; Doerr, W.W. *Chem. Eng. Prog.* May, 1998, p 19.
4. McKinnes, R.G. *Chem. Eng. Prog.* November 1995, p 36.
5. Dempsey, C. R.; Oppelt E.R. *Air Waste* **1993**, *43*, 25.
6. Tahmasseb, L.A.; Nelieu, S.; Einhorn, K.J. *Sci. Total Environ.* **2002**, *291*, 33.
7. Ollis, D.F.; Pelizzetti, E.; Serpone, N. *Environ. Sci. Technol.* **1991**, *25*, 1523.
8. Pulgarín, C.; Kiwi, J. *Chimia* **1996**, *50*, 50.
9. Andreozzi, R.; Caprio, V.; Insola, A.; Martota, R. *Cat. Today* **1999**, *53*, 51.
10. Chiron, S.; Fernández-Alba, A.; Rodríguez, A.; García-Calvo, E. *Wat. Res.* **2000**, *34*, 366.
11. Legrini, O.; Oliveros, E.; Braun, A.M. *Chem. Rev.* **1993**, *93*, 671.
12. Bauer, R.; Waldner, G.; Fallmann, H.; Hager, S.; Klare, M.; Krutzler, T.; Malato, S.; Maletzky, P. *Cat. Today* **1999**, *53*, 131.
13. Blanco, J.; Malato, S. *Solar Detoxification*. World Solar Progr. 1996-2005; Natural Sciences; UNESCO, 2001. (http://www.unesco.org/science/wsp)
14. Blanco, J.; Malato S. In *Water Recycling and Resource Recovery in Industry: Analysis, Technologies and Implementation*; Lens, P; Hulshoff, L.; Wildener, P.; Asano, T., Eds.; IWA Pub.: London, UK, 2002, Chapter 30.
15. Robert, D.; Malato, S. *Sci. Total Environ.* **2002**, *291*, 85.
16. Alfano, O.M.; Bahnemann, D.; Cassano, A.E.; Dillert, D.; Goslich, R. *Cat. Today* **2000**, *58*, 199.
17. Malato, S.; Blanco, J.; Vidal, A.; Richter, C. *Appl. Catal. B: Environ.* **2002**, *37*, 1.
18. Pignatello, J.J. *Environ. Sci. Technol.* **1992**, *26*, 944.
19. Fenton, H.J.H, *J. Chem. Soc.* **1894**, *65*, 899.
20. Sagawe, G.; Lehnard, A.; Lubber, M.; Bahnemann, D. *Helvetica Chimica Acta* **2001**, *84*, 3742.
21. Blake D.M. *Bibliography of Work on the Photocatalytic Removal of Hazardous Compounds from Water and Air*; National Technical Information Service, US Dep. of Commerce, Springfield, VA, USA, 2001.
22. Cunningham, J.; Al-Shayyed, G.; Sedlak, P.; Caffrey, J. *Cat. Today* **1999**, *53*, 145.
23. Ajona, J. A.; Vidal, A. *Sol. En.* **2000**, *68*, 109.
24. Muschaweck, J.; Spirkl, W.; Timinger, A.; Benz, N.; Dörfler, M.; Gut, M.; Kose, E. *Solar Energy* **2000**, *68*, 151.
25. Tripanagnostopoulos, Y.; Yianoulis, P.; Papaefthimiou, S.; Souliotis, M.; Nousia, Th. *Renewable Energy* **1999**, *16*, 628
26. Maurino, W; Minero, C.; Pelizzetti, E. *Chim. Ind.* **1999**, *81*, 61.

27. Minero, C.; Pelizzetti, E.; Malato, S.; Blanco, J. *So. En.* **1996**, *56*, 411-419.
28. Herrmann, J.M.; Disdier, J.; Pichat, P.; Malato, S.; Blanco, J. *Appl. Catal. B: Environ.* **1998**, *17*, 15.
29. Agüera, A.; Almansa, E.; Malato, S.; Maldonado, I.; Fernandez-Alba, A.; *Analuses* **1998**, *26*, 245.
30. Malato, S.; Blanco, J.; Richter, C.; Milow, B.; Maldonado, M.I. *Chemosphere* **1999**, *38*, 1145.
31. Fallmann, H.; Krutzler, T.; Bauer, R.; Malato, S.; Blanco, J. *Cat. Today* **1999**, *54*, 309.
32. Malato, S.; Blanco, J.; Fernandez-Alba, A. R.; Agüera, A. *Chemosphere* **2000**, *40*, 403.
33. Malato, S.; Blanco, J.; Richter, C.; Maldonado, M. I. *Appl. Catal. B: Environ.* **2000**, *25*, 31.
34. Malato, S.; Blanco, J.; Richter, C.; Fernández, P.; Maldonado, M. I. *Sol. En. Mat. Sol. Cells* **2000**, *64*, 1.
35. Agüera, A.; Almansa, E.; Tejedor, A.; Fernández-Alba, A.R.; Malato, S.; Maldonado, M.I. *Env. Sci. Technol.* **2000**, *34*, 1563.
36. Parra, S.; Sarriá, V.; Pulgarín, C.; Malato, S.; Peringer, P. *Appl. Catal. B: Environ.* **2000**, *27*, 153.
37. Malato, S.; Blanco, J.; Maldonado, M. I.; Fernández-Ibáñez, P.; Campos, A.; *Appl. Catal. B: Environ.* **2000**, *28*, 163.
38. Malato, S.; Caceres, J.; Aguera, A.; Mezcua, M.; Hernando, D.; Vial, J.; Fernández-Alba, A. R. *Env. Sci. Technol.* **2001**, *35*, 4359.
39. Parra, S.; Pulgarín, C.; Malato, S. *Appl. Catal. B: Environ.* **2002**, *36*, 131.
40. Malato, S.; Blanco, J.; Cáceres, J.; Fernández-Alba, A. R.; Agüera, A.; Rodríguez, A. *Cat Today* **2002**, in press.
41. Fernández-Alba, A. R.; Hernando, D.; Agüera, A.; Cáceres, J.; Malato, S. *Wat. Res.* **2002**, in press.
42. Parent, Y.; Blake, D.; Magrini-Bair, K.; Lyons, C.; Turchi, C.; Watt, A.; Wolfrum, E.; Prairie, M. *Sol. En.* **1996**, *56*, 429.
43. Malato, S.; Maldonado, M.I.; Blanco, J. *Descontaminación de Aguas De Lavado de Plaguicidas mediante Fotocatálisis Solar*; CIEMAT: Madrid, Spain. 2001.
44. Romero, M.; Blanco, J.; Sánchez, B.; Vidal, A.; Malato, S.; Cardona, A.; García, E. *Sol. En.* **1999**, *66*, 169.
45. Herrmann, J.M. *Cat. Today* **1999**, *53*, 115.
46. Dillert, R.; Cassano, A. E.; Goslich, R.; Bahnemann, D. *Cat. Today* **1999**, *54*, 267.
47. Brown, R.A.; Nelson, C.; Leahy, M. *In Situ and on-Site Bioremediation;* Batelle press: Columbus, OH, 1997, pp 457-462.
48. Sarriá, V.; Parra, S.; Adler, N.; Péringer, P.; Benitez N.; Pulgarin C. *Cat. Today* **2002**, in press.
49. Fernández-Alba, A.R.; Hernando, L.; Díaz, G.; Chisti, Y. *Anal. Chim. Acta* **2001**, *426*, 289.
50. Tothill, I.E.; Turner, A.P.F. *Trends Anal. Chem.* **1996**, *15*, 178.
51. Valming, V.; Connor, V.; Digiorgio C.; Bailey, H.C.; Deanovic, L.A.; Hinton D.E. *Environ. Tox. Chem.* **2000**, *19*, 42.

Field Processes and Applications

Chapter 10

Microbial Degradation of Atrazine in Soils, Sediments, and Surface Water

Mark Radosevich[1] and Olli H. Tuovinen[2]

[1]Biosystems Engineering and Environmental Science, University of Tennessee, 2506 E.J. Chapman Drive, Knoxville, TN 37996–4531
[2]Department of Microbiology, Ohio State University, 484 West 12th Avenue, Columbus, OH 43210–1292

The purpose of this review is to examine linkages between atrazine degradation potential, microbial ecology and detection of known catabolic genes in soil and aquatic environments with varying exposure to s-triazines. The biodegradation of atrazine and other chloro-s-triazine herbicides generally involves a series of hydrolytic reactions catalyzed by enzymes in the amidohydrolase superfamily. The major enzymatic steps, first identified in *Pseudomonas* sp. ADP and common to some other Gram-negative atrazine-mineralizing bacteria are dehalogenation, N-dealkylation of ring substituents (hydrolytic removal of ethylamino and isopropylamino moieties), and ring-cleavage of cyanuric acid, a central metabolite in the degradation of all s-triazines. The genes *atzABC* that encode the enzymes leading to production of cyanuric acid are plasmid borne in *Pseudomonas* ADP and several other bacteria and are widely distributed in Gram-negative atrazine-degraders. The role of these bacteria and catabolic genes they possess is unclear in natural environments.

Atrazine and other members of the *s*-triazine family of herbicides are widely used in weed control and consequently often detected in environmental media. Atrazine can act as an endocrine disruptor at environmentally relevant concentrations (0.1-25 ppb), resulting in severe developmental abnormalities in the reproductive system of amphibians (1,2). Thus, understanding the factors governing the persistence and bioavailability of these compounds is vitally important. Although atrazine is volatilized and transformed abiotically in minor amounts (3), the degradation by bacteria and fungi is the primary mode of attenuation in the environment. Atrazine is rendered less bioavailable through a variety of sorption and sequestration reactions in soils and sediments (4-8). This paper briefly summarizes what is known regarding microbial degradation of atrazine and examines recent findings in the biodegradation and mineralization of atrazine in terrestrial and aquatic environments.

Microbial Degradation of *s*-Triazines

The number of bacteria and eukaryotes known to partially transform or completely mineralize *s*-triazines has grown substantially in the past decade. These cultures have generally been obtained from soils and sediments through enrichment techniques. Enrichment conditions selecting for microorganisms capable of utilizing triazine-N as the sole source of N for assimilation have been most successful, but many organisms capable of using atrazine as the sole source of C and energy have also been isolated. While some variation does exist, the basic pathway in Gram-negative bacteria known to mineralize atrazine consists of upper (dehalogenation and sequential *N*-dealkylation leading to cyanuric acid) and lower (hydrolytic ring-cleavage of cyanuric acid ultimately yielding CO_2 and NH_4^+) pathways (Figure 1 and 2 below). The first pure bacterial isolates with complete atrazine mineralization were reported in the mid-1990's (9-11). Since that time, many bacteria have been isolated that can mineralize atrazine via upper and lower pathways shown in Figure 1 and 2 (12, 13). Although a great deal has been learned from the ever-growing list of atrazine-degrading bacteria isolated from soils and sediments, it remains unclear, what role if any these organisms and their catabolic genes play in the dissipation of *s*-triazine herbicides from natural environments. As a starting point for linking atrazine degradation potential in soils and sediments to known catabolic genes and specific bacterial groups that mineralize atrazine recent investigations have focused on the initial hydrolytic steps, dechlorination and ring-cleavage in the upper and lower pathways, respectively.

Biologically mediated dehalogenation of atrazine to 2-hydroxyatrazine by a bacterial enrichment culture was first shown by Mandelbaum et al. (14). Subsequently, Mandelbaum et al. (10) reported dehalogenation of atrazine by a pure bacterial culture (*Pseudomonas* ADP), and the gene encoding the atrazine dehalogenation activity has been cloned and characterized (15). The atrazine chlorohydrolase gene, *atzA*, hybridized with genomic DNA from geographically and phylogenetically diverse atrazine-degrading isolates from the U.S.A., Asia, and Europe (12). All evidence published thus far suggests that biologically mediated atrazine hydrolysis is widespread and contributes to the formation of hydroxyatrazine in natural environments. Thus many have assumed that the presence of the *atzA* gene in soils may be a useful index of atrazine degradation potential.

In the lower pathway, ring cleavage of cyanuric acid releases biuret and CO_2 (Figure 2). Biuret is further hydrolyzed to urea and then converted through urease activity to NH_4^+ and CO_2. Ring-cleavage is catalyzed by cyanuric acid amidohydrolase, encoded by *trzD* homologs in atrazine-mineralizing bacteria

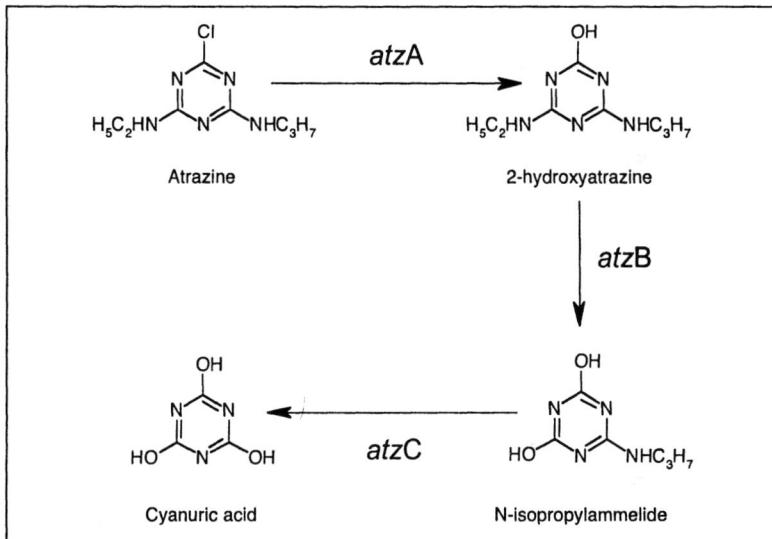

Figure 1. Upper atrazine biodegradation pathway in all known atrazine-mineralizing bacteria showing hydrolytic conversion of atrazine to cyanuric acid. Dehalogenation of atrazine to 2-hydroxyatrazine is catalyzed by the gene product of *atzA*. 2-Hydroxyatrazine is then sequentially *N*-dealkylated to produce cyanuric acid by enzymes encoded by *atzB* and *atzC*.

including a soil bacterial isolate M91-3 (16). This bacterium was recently classified as *Ralstonia basiliensis* (17). The enzyme from M91-3 was purified and the products of the reaction, ^{13}C-cyanuric acid → ^{13}C-biuret and $^{13}CO_2$, were confirmed by ^{13}C-NMR analysis (16). Further characterization of purified cyanuric acid amidohydrolase from *R. basilensis* M91-3 was completed to define the optimum pH and temperature, substrate specificity, and kinetic

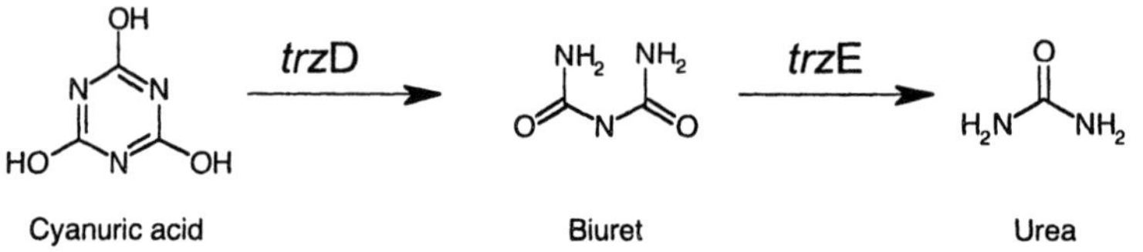

Figure 2. Lower s-triazine biodegradation pathway. Ring-cleavage is catalayzed by cyanuric acid amidohydrolase encoded by trzD, followed by biuret hydrolase, trzE.

parameters. The N-terminal sequence of cyanuric acid amidohydrolase had 100% homology to the N-terminus and to an internal sequence of the ring-cleavage enzyme (trzD) from *Pseudomonas* NRRLB-12227, which was originally characterized by Karns (18).

Atrazine Degradation Potential in Natural Environments

Soil. The atrazine history effect, leading to accelerated biodegradation, is well documented in agricultural soils. The rates of atrazine biodegradation and mineralization generally increase in soils with a history of atrazine exposure (19-21). The acclimation process has not been characterized at the microbial community level and the environmental and biological factors underlying this history effect remain obscure. Repeated applications of atrazine may enrich for microbial populations with the capacity to utilize the chemical as a carbon and energy source and thus an increase in the numerical abundance of that population would be anticipated. Such correlations have been documented for 2,4-dichlorophenoxy acetate, 2-methyl-4-chlorophenoxyacetate, EPTC, and carbofuran (22-26). It is possible that low concentrations of residual atrazine and metabolites in soils serve to keep the pathway enzymes expressed.

Atrazine is a relatively poor source of C and energy but potentially an excellent N-source (5 mol N per 1 mol atrazine). A positive association between the mineralization of [^{14}C-*ethyl*]-atrazine and the most probable number (MPN) of atrazine-degrading bacteria was observed under C-limiting conditions, but the relationship was not clear for ring-cleaving microorganisms (27). Ostrofsky et al. (28, 29) examined the history effect using cyanuric acid and atrazine as substrates in laboratory incubation experiments. Although atrazine mineralization was enhanced with prior exposure, the results on the increase in the abundance of atrazine-degraders were inconclusive. The most-probable-number (MPN) methodology was deemed to be inadequate in recovering atrazine-degraders quantitatively from agricultural soil samples. DNA was extracted from the MPN tubes and probed with *atzA* and *trzD*. However, there was no clear nor consistent association between the MPN counts and dot blot signals even in samples that actively mineralized atrazine (29). These data may indicate that these two genes did not represent the pathway of atrazine degradation in the soil samples examined.

To characterize the atrazine acclimation process at the microbial community level, samples of soils with (H-soil) and without (NH-soil) history of s-triazine application were collected from experimental plots on the Delaware Agricultural Experiment Station, Newark DE (30). H-soil samples were collected from continuous corn plots used in the corn breeding program. These plots received split annual application of herbicides containing atrazine at the maximum labeled rate. The NH-soils were collected from plots nearby that had been used for soybean research but were fallow at the time of sampling. These soils had no recent history of atrazine application. The samples were amended three times with atrazine and alternative nutrient sources at two-week intervals. Two culture-independent methods, fatty acid methyl ester analysis (FAME) and denaturing gradient gel electrophoresis (DGGE), were used to examine the changes in microbial community structure in response to atrazine application and nutrient availability.

Repeated application of atrazine enhanced the mineralization in the H-soil and NH-soil samples (30). The effect was most pronounced in the NH-soil. A high C/N ratio in soil also favored atrazine mineralization prior to acclimation. The addition of ammonium nitrate strongly inhibited the mineralization of atrazine in the NH-soil especially. MPN estimates for atrazine-degrading bacteria ranged from below detection to 10^7 cells g^{-1} soil (Table 1). An association between MPN values and the rate of atrazine mineralization could not be discerned because some soils with low MPN values (< 100 cells g^{-1} soil) had some of the highest atrazine mineralization rates after acclimation. The cell counts based on the MPN estimates in the H-soil were relatively high upon repeated atrazine exposure. In some NH-soil treatments, a delayed increase in cell numbers was observed 12 weeks after the final amendment of atrazine. It was concluded that enhanced atrazine mineralization was not correlated with the density of culturable atrazine-degrading bacteria. Changes in the community structures, as analyzed by DGGE and FAME, corresponded well with mineralization patterns. These data suggested that distinctly different atrazine-mineralizing populations developed as a function of C/N ratio and atrazine exposure in the soils examined. A significant observation revealed by DGGE analysis was a very prominent band (arrow) that appeared in atrazine amended soil (Figure 3). The band intensified in atrazine treated soil amended with either acetate or pectin (high C/N treatments) yet was noticeably absent in all soils amended with inorganic N including those that also received atrazine. This pattern corresponded well with atrazine mineralization capacity (data not shown). Presumably the phylotype represented by this band in the DGGE gel play a very significant role in atrazine degradation in this soil. Sequence analysis revealed that this environmental sequence was most closely related to a

Bacillus sp. Interestingly, only two Gram-positive atrazine-degrading species, both of the genus *Arthrobacter* sp., have been previously identified (31, 32).

At the conclusion of the acclimation study, soils from the microcosms were used to inoculate enrichment media of the same composition as the treatments used in the acclimation study. The atrazine concentration was monitored by HPLC and the enrichments were subcultured when the atrazine concentration had declined by at least 25% relative to sterile controls. Enrichments were subcultured seven times, and then DNA was extracted from subsamples and subjected to PCR-DGGE analysis as in the acclimated soils. The DGGE profiles for soils and enrichment cultures differed. Sequencing results suggested that many phylotypes associated with atrazine treatments were not closely related to previously characterized atrazine-degrading bacterial isolates.

Based on these results, it appears that atrazine mineralization rates and the density of culturable atrazine-degrading bacteria are not linked; i.e., the exposure to atrazine stimulates the activity of a population that was present in soil but was not detectable by the MPN assay.

In the soils and sediments examined, there appeared to be no uniform association between the frequency of occurrence of $atzA$, kinetics of mineralization, and MPN estimates of atrazine-degrading bacteria (30).

Aquatic and Wetland Systems. Relatively few studies have examined atrazine degradation potential in aquatic and wetland ecosystems. Ro and Chung (33) measured atrazine biodegradation potential in spiked wetland sediments that received sugar-mill waste water. Although mineralization of atrazine was not measured, degradation as determined by HPLC was observed and the rate increased with subsequent additions of atrazine. In another study in wetlands receiving agricultural runoff, atrazine was degraded in both the sediments and water column (34). Mesocosms were used to demonstrate accelerated atrazine degradation in sediments amended with sucrose. Anderson et al. (35) reported on aerobic mineralization of atrazine in sediment and water samples from a constructed wetland system fed by the Olentangy River (central Ohio) that drains an agricultural watershed. The river water also contained atrazine-mineralizing bacteria, perhaps originating from the agricultural fields in the watershed. In the constructed wetland site, the rate of mineralization was highest in the surface section (0-5 cm depth) of the sediment. Atrazine mineralization in the wetland was, therefore, attributed to active microorganisms in the river inflow and microbial populations already residing in the sediment zones. Atrazine mineralization was not detected in a groundwater-fed wetland, Cedar Bog (central Ohio), despite being surrounded by fields that had recently been treated with atrazine (35). Larsen et al. (36) reported the lack of atrazine mineralization under various redox regimes that ranged from fully aerobic to highly reducing conditions, suggesting that the microbial community in this ecosystem lacked the catabolic potential to degrade atrazine.

Table 1. Abundance of atrazine degrading bacteria in laboratory acclimated soil microcosms two weeks following the final amendment cycle. A five tube most probable number technique was used to estimate the number of culturable atrazine-degrading bacteria. Positive tubes were scored based on 25% loss of atrazine relative to sterile control tubes (30).

Treatment	Soil	
	History	Non-History
	Log cells g^{-1}	Log cells g^{-1}
Prior to acclimation	5.3	2.2
Microcosm Amendments		
Water	<2	<2
Acetone	<2	<2
NH$_4$NO$_3$	2.8	<2
Atrazine	5.0	2.7
Atrazine + NH$_4$NO$_3$	2.7	<2
Cyanuric Acid	2.3	<2
Atrazine + Cyanuric Acid	2.5	<2
Atrazine + Cyanuric Acid + NH$_4$NO$_3$	4.6	4.5
Atrazine + Acetate	3.2	3.6
Atrazine + Acetate + NH$_4$NO$_3$	2.3	3.2
Atrazine + Pectin	4.8	4.5
Atrazine + Pectin + NH$_4$NO$_3$	2.5	<2

Phylogenetics of Known Atrazine Metabolizing Bacteria

Previously described atrazine-degrading Gram-negative bacteria include *Pseudomonas*, *Pseudoaminobacter*, *Ralstonia*, *Alcaligenes*, *Chelatobacter*, *Aminobacter*, *Stenotrophomonas*, and *Agrobacterium* spp. All atrazine-mineralizing isolates within these genera appear to possess *atzABC* homologs encoding enzymes that produce cyanuric acid. They may also have either *atzD* or *trzD* homologs that encode cyanuric acid amidohydrolase for ring-cleavage. A Gram-positive organism, *Arthrobacter crystallopoides*, isolated from atrazine-treated agricultural soil, produced cyanuric acid from atrazine but possessed only *atzBC* as demonstrated by Southern hybridization using *atzABC* as probes (31). Only a weak signal was detected with *atzA*, suggesting that the bacteria contained an *atzA* homolog but with low extent of similarity in the nucleotide sequence. Similar results were very recently reported by Strong et al. (32) for an

Arthrobacter aurescens TC1 strain that was isolated by direct plating of contaminated soil on atrazine fortified agar plates. These results may suggest that selective enrichment in liquid media preferentially enriches for Gram-negative atrazine-degrading bacteria with catabolic pathways consisting of *atzABC*.

Atrazine Catabolic Genes in the Environment

Relatively little is known about the presence of catabolic genes in natural atrazine-degrading microbial communities. Shapir et al. (37), using a nested PCR approach, reported a positive association between the *atzA* copy number and the capacity of atrazine mineralization in agricultural soils. The *atzA* copy number ranged from $<1 \times 10^2$ to 1×10^5 copies g^{-1} soil in non-acclimated and acclimated soils, respectively suggesting that the presence of *atzA* was a good index of atrazine mineralization capacity.

In a recent study of atrazine degradation in wetlands sediments, an attempt was made to link the frequency of *atzA* and *trzD* to the atrazine mineralization capacity in the Cedar Bog (a spring-fed wetland) and Olentangy River Research Wetland Park (35). Relatively rapid mineralization of atrazine was observed in sediment samples (0-5 cm) collected from the Olentangy wetland. Mineralization was also high in the water column but only after the microbial community was concentrated (approximately 200-fold) before measurement of the activity. PCR amplification of genomic DNA extracted from the wetland sediment samples showed no positive signals for the *atzA* gene in Olentangy wetland samples, whereas Southern blots of the amplified DNA showed positive signals for some of them. The PCR data suggested that *trzD* was readily amplified from DNA extracted from the Olentangy wetland samples. Mineralization was negligible in Cedar Bog samples, and neither *atzA* nor *trzD* was detected. These results suggest that the frequency of *atzA* in these soils was low; yet some soils typically displayed high capacity of atrazine mineralization. It is plausible that there are other genes that are central in atrazine degradative pathways, such as the recently identified *trzN* (atrazine chlorohydrolase) from a *Nocardioides* isolate (38). Thus the possibility exists that the genes involved in the biodegradation of atrazine are diverse in soil microorganisms and do not yield signals when probed with genes isolated from known pure cultures.

Conclusions

Atrazine is a xenobiotic molecule and present in the environment mostly at ppb concentrations except for the couple of months of low ppm levels following its preemergence application on an annual cycle (excluding spill sites). Analytical data suggest that atrazine residues are never completely removed from soil environments during the annual cycle, and it remains unclear how efficient the biodegradation of atrazine is in scavenging the compound at low ppb concentrations such as those found in history soils. Rates of atrazine degradation vary widely in soils, aquifer sediments, and aquatic environments. While physical and chemical processes such as sorption can often influence

137

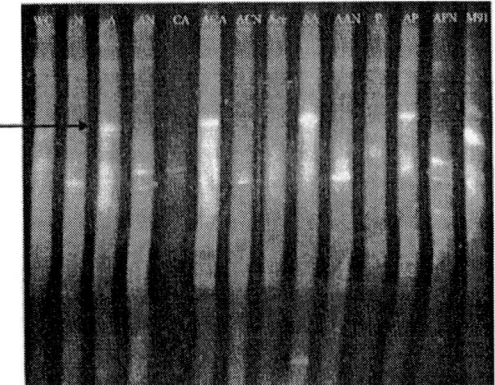

Figure 3. PCR-DGGE of whole soil community 16S rDNA The soils were amended as follows: WC, water control; N; NH_4NO_3; A, atrazine; AN, atrazine + NH_4NO_3; CA, cyanuric acid; ACA, atrazine + cyanuric acid; ACN, atrazine + cyanuric acid + NH_4NO_3; Ace, acetate; AA, atrazine + acetate; AAN, atrazine + acetate + NH_4NO_3; P, pectin; AP, atrazine + pectin; APN, atrazine + pectin + NH_4NO_3; M91, *Ralstonia basilensis*. Arrow indicates prominent band in atrazine-amended soils that did not receive ammonium nitrate.

degradation, the presence of an acclimated/adapted microbial community with the catabolic capacity to degrade atrazine is prerequisite for rapid dissipation from the environment. It is clear from a number of recent studies that enhanced or accelerated rates of atrazine degradation occur when environmental media are repeatedly exposed to atrazine suggesting an alteration in the structure and function of exposed microbial communities. The detection of *atzA* and *trzD* in soils that contain active atrazine-mineralizing microbial populations is variable, suggesting that there is substantial heterogeneity in the pathway genes to be discovered. The *trzN* gene is an example of a pathway segment that has not been probed in natural microbial populations. This variability suggests that the pathway genes do not have common ancestry but have evolved into pathways in a mosaic fashion in different microorganisms. Thus, the current collection of bacterial isolates that mineralize atrazine may not adequately reflect the extant phylogenic or catabolic diversity of atrazine-degrading communities in natural environments.

Acknowledgments.

Partial funding for this work was received from the U.S. Department of Agriculture (Grant No. 98-35107-6388). We are grateful to K.L. Anderson,

E.B. Ostrofsky, E.D. Rhine, J.B. Robinson, D.M. Stamper, and K.A. Wheeler for their past contributions in our research group.

References

1. Hayes, T.B.; Collins, A.; Lee, M.; Mendoza, M.; Noriega, N.; Stuart, A.A.; Vonk, A. *Proc. Natl. Acad. Sci. USA,* **2002a**, 99:5476.
2. Hayes, T.; Haston, K.; Tsui, M.; Hoang, A.; Haeffele, C.; Vonk, A. *Nature,* **2002b**, 419:895.
3. Weber, J.B.; Hardy, D.H.; Leidy, R.B. In: *Pesticide environmental fate: bridging the gap between laboratory and field studies*; Phelps, W.; Winton, K.; Effland, W.R., Eds., American Chemical Society, Washington, D.C. 2002; pp. 125-141.
4. Welhouse, G.J.; Beam, W.F. *Environ. Sci. Technol.,* **1993a**, 27:494.
5. Welhouse, G.J.; Beam, W.F *Environ. Sci. Technol.,* **1993a**, 27:500.
6. Mersie, W.; Liu, J.; Siebold, C.; Tierney, D. *Weed Sci.,* **1998**, 46:480.
7. Houot, S.; Topp, E.; Yassir, A.; Soulas, G. *Soil Biol. Biochem.,* **2000**, 32:615.
8. Munier-Lamy, C.; Feuvrier, M.P.; Choné, T. *J. Environ. Qual.,* **2002**, 31:241.
9. Yanze-Kontchou, C.; Gschwind, N. *Appl. Environ. Microbiol.,* **1994**, 60:4297.
10. Mandelbaum, R.T.; Allan, D.L.; Wackett, L.P. *Appl. Environ. Microbiol.* **1995**, 61:1451.
11. Radosevich, M.; Traina, S.J.; Hao, Y.-L.; Tuovinen, O.H. *Appl. Environ. Microbiol.,***1995**, 61:297.
12. Sadowsky, M.J.; Wackett, L.P. In: *Pesticide biotransformation in plants and microorganisms: similarities and divergences*; Hall, J.C.; Hoagland, R.E.; Zablotowicz, R.M., Eds.; American Chemical Society: Washington, D.C., 2001; pp. 268-282.
13. Wackett, L.P.; Sadowsky, M.J.; Martinez, B.; Shapir, N. *Appl. Microbiol. Biotechnol.,* **2002**, 58:39.
14. Mandelbaum, R.T.; Wackett, L.P.; Allan,D.L. *Appl. Environ. Microbiol.,* **1993**, 59:1695.
15. de Souza, M.L.; Wackett, L.P.; Boundy-Mills, K.L.; Mandelbaum, R.T.; Sadowsky, M.J. *Appl. Environ. Microbiol.* **1995**, 61:3373.
16. Stamper, D.M. 2000. Ph.D. Thesis, Ohio State University, Columbus, OH, 43210.
17. Stamper, D.M.; Radosevich, M.; Hallberg, K.B.; Traina, S.J.; Tuovinen, O.H. *Can. J. Microbiol* **2003**, in press.
18. Karns, J.S. *Appl. Environ. Microbiol.,* **1999**, 65:3512.
19. Ostrofsky, E.B.; Traina, S.J.; Tuovinen, O.H. *J. Environ. Qual,* **1997**, 26:64.
20. Vanderheyden, V.; Debongnie, P.; Pussemier, L. *Pestic. Sci.,* **1997**, 49:237.
21. Yassir, A.; Lagacherie, B.; Houot, S.; Soulas, G. *Pestic. Sci.,* **1999**, 55:799.
22. Ka, J.O.; Burauel, P.; Bronson, J.A.; Holben, W.E.; Tiedje, J.M. *Soil Sci. Soc. Am. J.,* **1995**, 59:1581.
23. Smith, A.E.; Aubin, A.J. *Bull. Environ. Contam. Toxicol.,* **1994**, 53:7.
24. Wilson, R.G. *Weed Sci.,* **1984**, 32:264.

25. Karpouzas, D.G.; Walker, A.; Froud-Williams, R.J.; Drennan, D.S.H. *Pestic. Sci.*, **1999**, 55:301.
26. Kotoula-Syka, E.; Hatzios, K.K.; Berry, D.F.; Wilson, H.P. *Weed Technol.*, **1997**, 11:403.
27. Jayachandran, K.; Stolpe, N.B.; Moorman, T.B.; Shea, P.J. *Soil Biol. Biochem.*, **1998**, 30:523.
28. Ostrofsky, E.B.; Robinson, J.B.; Traina, S.J.; Tuovinen, O.H. *Soil Biol. Biochem.*, **2001**, 33:1539.
29. Ostrofsky, E.B.; Robinson, J.B.; Traina, S.J.; Tuovinen, O.H. *Soil Biol. Biochem.*, **2002**, 34:1449.
30. Rhine, E.D. **2001**, Ph.D. Thesis, University of Delaware, Newark, DE, 19716.
31. Rousseaux, S.; Hartmann, A.; Soulas, G. *FEMS Microbiol. Ecol.*, **2001**, 36:211.
32. Strong, L.C.; Rosendahl, C.; Johnson, G.; Sadowsky, M.J.; Wackett L.P. *Appl. Envir. Microbiol.*, **2002**, 68: 5973.
33. Ro, K. S.; Chung, K. H. *J. Environ. Sci. Health A*, **1995**, 30: 121.
34. Kao, C.M.; Wang, J.Y.; Wu, M.J. *Wat. Sci. Technol.*, **2001**, 44:539.
35. Anderson, K.L.; Wheeler, K.A.; Robinson, J.B.; Tuovinen, O.H. *Water Res.*, **2002**, 36:4785.
36. Larsen, L.; Jørgensen, C.; Aamand, J. *J. Environ. Qual.* **2001**, 30:24.
37. Shapir, N.; Goux, S.; Mandelbaum, R.T.; Pussemier, L. *Can. J. Microbiol.*, **2000**, 46:425.
38. Mulbry, W.W.; Zhu, H.; Nour, S.M.; Topp, E. *FEMS Microbiol. Lett.*, **2002**, 206: 75.

Chapter 11

Bioremediation of Atrazine-Contaminated Soil

Edward Topp[1], Fabrice Martin-Laurent[2], Alain Hartmann[2], and Guy Soulas[2]

[1]Agriculture and Agri-Food Canada, 1391 Sandford Street, London, Ontario N5V 4T3, Canada
[2]INRA Dijon, 17 Rue Sully, 21065 Dijon, France

> The widely-used herbicide atrazine poses a threat to environmental health if point sources due to spills are not efficiently dealt with. Numerous atrazine-degrading microorganisms have been been isolated, and there is much known about the genetics and biochemistry of atrazine biodegradation. A variety of studies have explored the utility of atrazine-degrading bacteria for soil bioremediation. The effects of various rate controlling factors, pH, availability of nutrients, terminal electron acceptor and efficacy of different bacteria have been explored. Finally, the advent of molecular methods used to monitor the persistence and assess the activity of bioremediating bacteria is likely to improve the success of bioremediation.

Introduction

Atrazine (2-chloro-4-ethylamino-6-isopropylamino-s-triazine) is one of the world's most heavily used agricultural herbicides. It controls a variety of broadleaf weeds infesting corn, sorghum and certain other crops. Atrazine is a moderately persistent herbicide, with half-lives ranging from several days to several months (1-4). It is frequently detected in surface water (5,6), rainwater (7), tile drainage (8) and groundwater (9, 10), frequently at concentrations exceeding the European Union standard of $0.1 \mu g\ l^{-1}$, and the US-EPA maximum contaminant level of 3µg l⁻¹ (11). Atrazine may affect sex differentiation in amphibians, reptiles and mammals by inducing aromatase, the enzyme that converts androgens to estrogens, and there is some evidence that amphibians in agricultural areas receiving atrazine have unusually frequent developmental abnormalities (12).

The environmental persistence, relatively high water-solubility, and leaching potential of this chemical have caused it to be banned in a number of jurisdictions, and it's use is restricted in others (4).

There is a significant body of information on the role of microbial biodegradation in mediating the persistence of this chemical in the environment, and the prospects of employing bioremediation to accelerate its destruction in cases where point sources pose an unacceptable environmental risk. This review will concentrate on the bioremediation of atrazine in soils by inoculation with atrazine-degrading bacteria, factors that influence the efficacy of this process, and the application of modern molecular techniques to probe the ecology and physiology of atrazine-degrading bacteria in soils. The reader is referred to recent reviews of the genetics and enzymology of atrazine-degrading bacteria (13), the history and use of triazine herbicides (14) and the general environmental behaviour of atrazine (4).

Mechanisms of Bacterial Degradation of Atrazine

Oxidative Attack

Deethylatrazine (CIAT) and deisopropylatrazine (CEAT) are frequently detected transformation products of atrazine in soil and water. Several bacteria and fungi which catalyze N-dealkylation of one or both side chains have been described (Table 1). A common feature of all of these organisms is that they do not cleave the *s*-triazine ring. Some actinomycete bacteria that employ monooxygenases to oxidatively N-dealkylate atrazine with the formation of deethylatrazine and deisopropylatrazine have been thoroughly studied. For example, *Rhodococcus* sp. strain NI86/21 has an inducible cytochrome P-450 enzyme system that catalyzes the dealkylation of atrazine (15). In this strain the genes *thc*B (coding for a novel type of cytochrome P-450 enzyme), *thc*C (coding for rhodocoxin) and *thc*D (coding for rhodocoxin reductase) together confer the atrazine-degrading phenotype. In the presence of oxygen *Rhodococcus* strain TE1 N-dealkylated atrazine and the other chlorotriazine herbicides propazine, simazine, and cyanazine using a pathway associated with a plasmid (16). Some actinomycete bacteria including *Streptomyces* sp. degrade atrazine by means of peroxidases, presumably fortuitously. For example *Streptomyces* sp. strainPS1/5 dealkylated atrazine when incubated in the presence of chitin (17). The white rot ligninolytic fungus *Phanerochaete chrisosporium* strain ATCC24725 degraded atrazine yielding mainly deethylatrazine but with some hydroxyatrazine production (18-20). The oxidative biodegradation of atrazine is likely to be an important means of dissipation in the environment. For bioremediation purposes these organisms are disadvantaged in that they do require oxygen, and that the dealkylated end products retain some phytotoxicity.

Hydrolytic Attack

Soils which have not been previously exposed to atrazine degrade the herbicide slowly and do not mineralize ^{14}C-ring labelled material (21). However, atrazine is now rapidly mineralized in North American and European agricultural soils that have repeatedly come in contact with the herbicide in normal farming practice (21,22). The ready isolation of numerous genera of gram-negative and gram-positive atrazine-degrading bacteria from these soils attests to the biological basis for this rapid degradation (Table 1; 23-29). Bacteria isolated from soils that rapidly mineralize the herbicide commonly initiate atrazine degradation by hydrolytic dechlorination, the product of which (hydroxyatrazine) is non-phytotoxic. The genes encoding an atrazine chlorohydrolase (*atz*A) and two amidohydrolytic reactions (*atz*B and *atz*C), which together convert atrazine to the ring-cleavage substrate cyanuric acid, have been cloned from *Pseudomonas* sp. strain ADP (30-33). Cyanuric acid is converted by another set of amidohydrolase enzymes (*atz*D, *atz*E, *atz*F) to biuret and urea, which are then mineralized (34, 35). The genes encoding these enzymes are widespread, highly conserved, plasmid-borne, and in some cases associated with catabolic transposons (28, 30, 36, 37). *Pseudomonas* sp. strain YAYA6 (DSM 93-99) was also found to mineralize atrazine by dechlorination and N-dealkylation (23). Isolates vary in the breadth of *s*-triazine herbicides degraded, for example Strain M91-3 (putatively identified as a *Ralstonia* sp.) mineralizes atrazine, simazine and cyanazine (38, 39). *Rhizobium* and *Agrobacterium radiobacter* strains were found to dechlorinate or mineralize atrazine, respectively (26, 27).

There is both commonality and variability in the genes encoding triazine hydrolases . For example *trz*N from *Nocardioides* sp. strain C190 was characterized recently, found to differ significantly from *atz*A in sequence, and to encode an enzyme with an *s*-triazine substrate specificity broader than that of *atz*A (28, 40). A comparison of different atrazine-degrading bacteria shows heterogeneity in the manner in which pathways are constructed, and the location

Table 1: Examples of atrazine-degrading microorganisms.

Microorganism	End products	Comments	Reference
Oxidative degradation			
Rhodococcus erythropolis. strain NI86/21	deisopropylatrazine, deethylatrazine, hydroxyisopropylatrazine	Cytochrome P-450 mediated oxidation (chromosomally encoded)	15
Rhodococcus sp strain TE1	deisopropylatrazine, deethylatrazine	Cytochrome P-450 mediated oxidation (plasmid encoded)	15, 16
Rhodococcus sp strain B30	deisopropylatrazine, deethylatrazine	Cytochrome P-450 mediated oxidation	15
Streptomyces sp. strainPS1/5	deisopropylatrazine, deethylatrazine	Peroxidases (?)	17
Phanerochaete chrysosporium strain ATCC24725	deethylatrazine hydroxyatrazine	Cytochrome P-450 mediated oxidation	18-20
Hydrolytic degradation			
Stable enrichment mixed culture	$CO_2 + NH_3$	Atrazine used as sole N source	69
Pseudomonas sp. strain YAYA6	$CO_2 + NH_3$, cyanuric acid		23
Pseudomonas sp. strain ADP	$CO_2 + NH_3$	*atz*ABCDE carried on plasmid pADP1	24, 30
Ralstonia sp. strain M91-3	$CO_2 + NH_3$		39
Rhizobium sp. strain PATR	hydroxyatrazine	*atz*A encoded atrazine chlorohydrolase	26
Agrobacterium radiobacter J14a	$CO_2 + NH_3$		27
Escherichia coli carrying pMD4	hydroxyatrazine	Recombinant killed cells expressing *atz*A	46

Pseudaminobacter sp. strain C147	$CO_2 + NH_3$	*atz*ABC	28
Nocardioides sp. strain C190	N-ethylammelide	*trz*N encoded triazine hydrolase	29, 40
Chelatobacter heintzii strain Cit1	$CO_2 + NH_3$	*atz*ABC, *trz*D	41, 42
Arthrobacter crystallopoiëtes Dij1	Cyanuric acid	*trz*N, *atz*BC	41, 42
Consortium consisting of 4 bacteria including *Clavibacter michiganese* strain ATZ1 and *Pseudomonas* sp. strain CN1	$CO_2 + NH_3$	*atz*ABC	31

of atrazine-degrading genes within the genome. For example, atrazine-degrading strains of the Gram negative genera *Chelatobacter heintzii, Aminobacter aminovorans, Stenotrophomonas maltophilia* and the Gram positive genus *Arthrobacter crystallopoïetes* vary in the end product of atrazine degradation (41). Most of the Gram negative bacteria mineralized [^{14}C] ring-labelled atrazine and carried the *atz*ABCD genes. The *Arthrobacter* converted atrazine to cyanuric acid, and carried only the *atz*B and *atz*C genes, *atz*A was not detected. In these strains genes involved in atrazine degradation were shown to be located on different plasmids (42). In the case of the Gram-negative strains of *Chelatobacter heintzii*, 6 to 7 plasmids were observed, the *atz*ABC and *trz*D genes were located on 2 or 3 plasmids with variable molecular weights. In *Arthrobacter crystallopoietes*, the *atz*BC genes were located on a single plasmid of 117 kb. In a collection of *Pseudaminobacter* sp. strains, the *atz*ABC genes, were not always clustered together on the same plasmid (28). Overall, there is significant diversity in the types of bacteria which hydrolytically dechlorinate atrazine (see Table 1), in the manner in which atrazine degrading genes may be recruited and organized to create atrazine metabolizing pathways, and in the end products of atrazine metabolism. This genetic diversity in atrazine-degrading bacteria offers significant potential for finding efficacious bioremediation agents.

Examples of Soil Bioremediation Using Atrazine-degrading Bacteria

Mineralization of the atrazine molecule in soil can be achieved though inoculation with an organism that carries out the complete pathway, or inoculation with mixed cultures whose members independently but sequentially carry out the various reactions in the complete mineralization pathway. *Nocardioides* Strain C190 converts atrazine to the end product N-ethylammelide. [*ring*-U-^{14}C]atrazine was mineralized after a lag phase in non-sterile but not in sterile loam soil containing 1 mg/kg atrazine inoculated with 10^5 CFU/gr of *Nocardioides* Strain C190. Both [*ring*-U-^{14}C]hydroxyatrazine and [*ring*-U-^{14}C]N-ethylammelide were readily mineralized when added to this soil at a concentration of 1 mg/kg (43). Ring-labelled atrazine added to a sandy-loam soil at an initial concentration of 6 mg/kg was rapidly mineralized upon inoculation with 4.5 X 10^6 cells/g of an atrazine-mineralizing consortium (44). The consortium consisted of a *Clavibacter* strain, a *Pseudomonas* strain, and at least two other uncharacterized members (31).The *Clavibacter* strain converted atrazine to N-ethylammelide, the *Pseudomonas* strain then dealkylated and mineralized the *s*-triazine ring. The *Clavibacter* could utilize either atrazine or isopropylamine as sole source of carbon and nitrogen. At higher initial atrazine concentrations of 100 mg/kg, repeated inoculations with 10^8 cells/g were required to achieve substantial mineralization of ring-labelled atrazine (45). In a large field scale mesocosm, 8 repeated inoculations with 10^5 cells/g of the consortium promoted the dissipation of 72% of an initial 100 mg/kg concentration of commercial atrazine (Aatrex 90) within 11 weeks. The atrazine-supplemented soil was aged for 3 weeks prior to the start of the experiment. No atrazine transformation products were detected by HPLC. Inoculum was added to the soil directly in spent culture medium, and therefore the carryover of nutrients may have contributed to the activity.

Bioremediation efficacy can be enhanced by the addition of a carbon source. Seventy per cent of an initial atrazine concentration of 1.45 g/kg soil was removed within 3 weeks following inoculation with 2 X 10^6 cells/g of *Pseudomonas* ADP with the simultaneous addition of 2 g/kg of sodium citrate (24). In this study, there was no significant effect of inoculation without citrate addition, but addition of sodium citrate on its own significantly stimulated dissipation, suggesting that there was a significant indigenous population of atrazine-degrading bacteria in this soil.

Microbial bioremediation systems can be designed which are completely independent of survival or growth of the bioremediating agent. A recombinant *E. coli* expressing *atz*A cloned from *Pseudomonas* ADP enhanced atrazine degradation in a heavily polluted field site when added as killed cells at 0.5% (w/w) (46). The *E. coli* engineered to produce atrazine chlorohydrolase was used to treat several hundred liters of soil excavated from a site in South Dakota contaminated with several grams of atrazine/ kg soil. The region containing *atz*A had previously been cloned from *Pseudomonas* ADP into a plasmid (pMD4) carried in *E. coli* DH5alpha, from which it was constitutively expressed (30). Cells were grown in a batch fermentor, harvested, and killed by treatment with gluteraldehyde. This treatment cross-linked atrazine chlorohydrolase, enhancing its stability during storage of the cells and following addition to the soil. The treatment killed the *E. coli* cells, satisfying regulatory concerns about the release of a viable recombinant bacterium. Incorporation of 5 g cells/kg soil effected a 53% reduction in the atrazine concentration, whereas incorporation of cells and addition of 300 mg/kg phosphate fertilizer effected a 77% reduction within 8 weeks. Atrazine concentrations were by comparison stable over this time in untreated soil, or soil that received only the phosphate treatment. Concentrations of hydroxyatrazine or other potential transformation products were not reported. In bench scale experiments with this soil, addition of various carbon sources was found to suppress atrazine degradation, pH adjustment to have no effect, and phosphate addition to be stimulatory. Low soil temperatures in the fall likely reduced the efficacy of the treatment. Overall, this study showed that delivery of a large amount of atrazine chlorohydrolase via a non-viable recombinant bacterium was an effective means to bioremediate large volumes of soil heavily contaminated with atrazine.

There have been relatively few controlled comparisons of the relative bioremediation efficacy of atrazine-degrading bacteria. Topp (43) compared the efficacy for bioremediation of *Pseudomonas* ADP, *Pseudaminobacter* C147, and *Nocardioides* C190. Resting cell suspensions of these three bacteria degrade atrazine at comparable specific activities. Strains ADP and C147 mineralize atrazine, whereas the end product of atrazine degradation by Strain C190 is N-ethylammelide. *Pseudomonas* ADP can utilize atrazine as sole N but not as sole C source, whereas Strains C147 and C190 can use it as both C and N source. When inoculated at 10^7 cfu/gr into a loam soil containing 1 mg atrazine /kg all three bacteria mineralized atrazine at comparable and rapid rates. However, when the inoculum density was 10^5 cfu/gr the *Pseudaminobacter* was effective, the *Pseudomonas* did not enhance mineralization, and the *Nocardioides* did enhance mineralization, but only after a long lag. A number of studies have shown that atrazine-degrading bacteria which are unable to use the compound as a carbon source can enhance atrazine bioremediation when added at relatively high inoculum levels, or when readily metabolizable carbon sources are provided to support growth. For example, at an inoculum density of 10^6 *Pseudomonas* strain ADP cells/g soil enhanced atrazine removal and mineralization when accompanied by addition of citrate (24, 47). A relatively heavy inoculum density of 10^8 cells/g enhanced atrazine mineralization, particularly when the inoculation was repeated at 3 day intervals (45). On the other hand, *Agrobacterium radiobacter* J14a, which can utilize the ethylamine sidechain carbon atoms, enhanced atrazine degradation when added to soil at 10^5 cells/g (27). Taken together, the ability to utilize the target pollutant as a carbon source can enhance efficacy of a bioremediation agent, and should therefore be an economically valuable trait, both with respect to the speed of remediation, and by reducing the amount of required inoculant. Pollutant-degrading bacteria have been isolated on the basis of their ability to utilize nitrogen, sulfur or phosphorus atoms in the target molecules (48) These results suggest that, where possible, it may be preferable to obtain potential bioremediating bacteria by using enrichment and isolation strategies that select for organisms that can use the target chemical as a carbon source.

Oxygen is not a reagent for any of the enzymatic reactions in the atrazine mineralization pathway initiated by hydrolytic dechlorination. Nevertheless, under anaerobic conditions obligate aerobes can't undertake the process efficiently because of their respiratory requirement for oxygen. In contrast, bioremediation by denitrifying atrazine-degrading bacteria is effective under anaerobic conditions if nitrate is available as an electron acceptor. Isolate M91-3 mineralized atrazine with simultaneous nitrate reduction in a fixed film reactor perfused with buffered nitrate-containing medium containing 21.6 mg atrazine/l as the sole carbon source (49, 50). Likewise, atrazine was completely degraded by a mixed culture in a fixed film reactor perfused at a hydraulic residence time of 3.6h with buffer containing nitrate and 7-15 mg atrazine /l (51).

Monitoring the Abundance and Activity of Bioremediating Bacteria in Soils

Methods to enumerate atrazine-degrading bacteria and monitor the expression of genes encoding atrazine-degrading enzymes hold great promise to explore soil-dependent factors that determine bioremediation efficacy (52). Atrazine-degrading microorganisms in soil can be enumerated radiorespirometrically using a ^{14}C-most-probable-number (MPN) method, diluting soil samples to extinction and estimating the population size on the basis of mineralization of chain- or ring-labeled ^{14}C-atrazine in an enrichment medium (53). This approach was used to study the impact of carbon and nitrogen amendment on atrazine-degrading populations in soils that varied in their ability to mineralize atrazine (54). Population sizes ranged widely from twenty to ten thousand atrazine-degrading cells/g. Both the rate of atrazine mineralization and the size of the atrazine-mineralizing populations were generally increased by the frequency of atrazine use, suggesting that exposure to the herbicide promoted the development of an atrazine-mineralizing population (54). The MPN method has recently been coupled with DNA hybridization to both enumerate atrazine-degrading microbial communities, and detect within them the presence of known genes (*atzA* atrazine chlorohydrolase from *Pseudomonas* sp.; ADP, *thcR and thcB* cytochrome P450 from *Rhodococcus* TE1; and *trzD* from *Pseudomonas* NRRLB-12228) encoding enzymes involved in atrazine degradation (55). The combination of fluorescent *in situ* hybridization (FISH) and micro-autoradiography may provide an interesting new tool for studying the microbial ecology of atrazine biodegradation (56). In this method, the incorporation of radioactive atoms into cells that metabolize a radiolabelled substrate is detected by microautoradiography, and the preparations are also analysed by FISH with probes that reveal bacterial identity (57). Microscopic examination of preparations thus treated simultaneously provides information on the number, the specific activity and the identity of atrazine-degrading microbes.

Within the last decade applications of new culture-independent molecular tools based on PCR analysis of soil-extracted nucleic acids have provided unique insights into the composition, richness and structure of microbial communities (58-61). Quantitative PCR methods will be used to estimate the abundance in soils of genes that encode atrazine-degradation enzymes, and by quantifying mRNA, their expression. Magnetic capture hybridization (MCH) followed by nested PCR could predict the potential of a soil to mineralize atrazine (62). In this study, *atzA* gene copy number was quantified and found to be correlated with the mineralization rate of atrazine. More recently, another quantitative PCR method, competitive PCR, was used to enumerate *atzA* gene copy number in soil (63). Using this approach, atrazine-degrading *Pseudomonas* sp. strain ADP were quantified in aquifer sediments. The authors concluded that quantitative PCR could be used as a tool for monitoring introduced atrazine-degrading microbes independent of culturability. In another study competitive PCR was used to estimate the abundance of *atzC* (64). It was shown using this technique that in response to atrazine treatment the amount of the gene *atzC* increased briefly in adapted soil pre-treated with atrazine, suggesting that accelerated atrazine biodegradation in soil could be due to a transient increase in the size of the atrazine mineralizing population. In addition, the application of *atzC* PCR on DNA samples extracted either from bulk soil or from maize rhizosphere showed a transient increase in the amount of *atzC* gene which was concomitant to atrazine mineralization (65). The recent development of 'real time' PCR methods will greatly increase the sensitivity, linear range, and ease of enumerating genes in soil DNA. This method was used to enumerate *trzN* for monitoring an atrazine-degrading *Nocardioides* sp. SP12 in either bulk soil or maize rhizosphere (66). The abundance of genes encoding atrazine-degrading enzymes is by no means the whole story, atrazine-degrading activity requires transcription of these genes to the corresponding RNA message (mRNA) used as the template for protein synthesis. Messenger RNA molecules are inherently shortlived, and obtaining stoichiometric preparations from soils that are sufficiently pure and stable for subsequent quantitative analysis represents a significant technical challenge. However, up to now a relatively small number of reports can be found in the literature concerning the detection and the quantification of mRNA translated from gene encoding functional microbial enzymes (67). In particular, this has not yet been reported for *atz* genes in soil, although it has recently been shown that *atz* genes are transcriptionally regulated in *Pseudomonas* sp. ADP (35, 68). Soil pH is a key factor that can influence the kinetics and pathways of atrazine degradation, soils with pH values inferior to 6.5 do not mineralize the herbicide (22). The effect of pH on expression of *atz*ABC was investigated in *Pseudomonas* sp. ADP (68). These studies revealed that (i) these genes are expressed at a basal level, (ii) their expression is enhanced in the presence of atrazine and (iii) that at pH values less than 6.5 their expression is not increased by exposure to atrazine (Fig 1). Together these data suggest that the effect of soil pH on the rate of atrazine mineralisation measured in the field could be due to an alteration of atrazine-degrading microbial functioning via the alteration of the level expression of atrazine degrading genes. From the bioremediation perspective, the challenge now is to find atrazine-degrading bacteria which do express their biodegradative enzymes under acidic soil conditions. Overall, this example illustrates how molecular tools to probe the physiology of atrazine-degrading bacteria in soils can yield insights into the mechanisms of rate-controlling factors, and help plan strategies to develop more robust bioremediating strains.

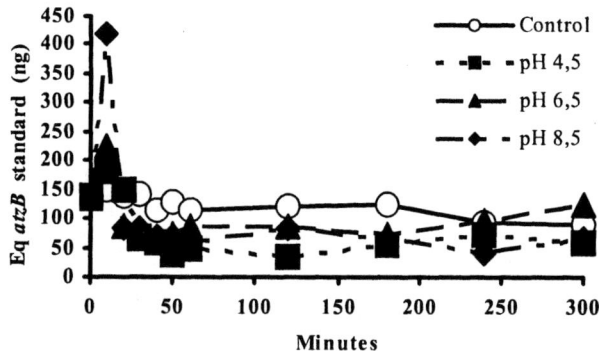

Figure 1. Expression of atzB in Pseudomonas ADP as a function of medium pH.

References

1) Muir, D.C.; Baker, B.E. *J. Agric. Food Chem.* **1976**, *24*, 122-125.
2) Dao, T.H.; Lavy, T.L.; Sorensen, R.C. *Soil Sci. Soc. Am. J.* **1979**, *43*, 129-1134.
3) Frank, R.; Sirons, G.J. *Bull. Environ. Contamin. Toxicol.* **1985**, *34*, 541-548.
4) Solomon, K.R.; Baker, D.B.; Richards, R.P.; Dixon, K.R.; Klaine, S.J.; Point, T.W.; Kendall, R.J.; Weisskopf, C.P.; Giddings, J.M.; Giesy, J.P.; Hall, L.W. Jr.; Williams, W.M. *Environ. Toxicol. Chem.* **1996**, *15*, 31-76.
5) Pereira, W.E. ; Rostad, C.E.; *Env. Sci. Technol.* **1990**, *24*, 1400-1406.
6) Spalding, R.F.; Snow, D.D.; Cassada, D.A.; Burbach, M.E. *J. Environ. Qual.* **1994**, *23*, 571-578.
7) Buser, H.R. *Environ. Sci. Technol.* **1990**, *24*, 1049-1058.
8) Buhler, D.D.; Randall, G.W.; Koskinen, W.C.; Wyse, D.L. *J. Environ. Qual.* **1993**, *22*, 583-588.
9) Wehtje, G.R.; Spalding, R.F.; Burnside, O.C.; Lowry, S.R. ; Leavitt, J.R.C. *Weed Sci.* **1983**, *31*, 610-618.
10) Pick, F.E.; van Dyk, L.P.; Botha, E. *Chemosphere.* **1992**, *25*, 335-341.
11) Thurman, E.M.; Goolsby, D.A.; Meyer, M.T.; Mills, M.S.; Pomes, M.L.; Kolpin, D.W. *Env. Sci. Technol.* **1992**, *26*, 2440-2447.

12) Hayes, T.B.; Collins, A.; Lee, M.; Mendoza, M.; Noriega, N.; Stuart, A.A.; Vonk, A. *Proc. Natl. Acad. Sci. USA.* **2002**, *99*, 5476-5480
13) Wackett, L.P. ; Sadowsky, M.J. ;Martinez, B. ; Shapir, N. *Appl. Microbiol. Biotehnol.* **2002**, *58*, 39-45.
14) Esser, H.O. ; Dupuis, G. ; Ebert, E. ; Marco, G.J. ; Vogel, C. In *Herbicides. Chemistry, degradation and Mode of Action* ; Editors Kearney, P.C., and Kaufman D.D. ;Marcel Dekker Inc. ; New York, NY ; **1975** ;pp 129-208
15) Nagy, I.; Compernolle, F.; Ghys, K.; Vanderleyden, J.; Demot, R. *Appl. Environ. Microbiol.* **1995**, *61*, 2056-2060.
16) Behki, R.; Topp, E.; Dick, W.; Germon, P. *Appl. Environ. Microbiol.* **1993**, *59*, 1955-1959.
17) Fadullon, F.S.; Karns, J.S.; Torrents, A. *J. Environ. Sci. Health Part B - Pesticides Food Contaminants and Agricultural Wastes* **1998**, *33*, 37-49.
18) Donnelly, P.K.; Entry, J.A.; Crawford, D.L. *Appl. Environ. Microbiol.* **1993**, *59*, 2642-2647.
19) Mougin, C.; Laugero, C.; Asther, M.; Dubroca, M.; Frasse, P.; Asther, M. *Appl. Environ. Microbiol.* **1994**, *60*, 705-708.
20) Mougin, C.; Laugero, C.; Asther, M.; Chaplain, V. *Pestic. Sci.* **1997**, *49*, 169-177.
21) Pussemier, L.; Goux, S.; Vanderheyden, V.; Debongnie, P.; Tresinie, I.; Foucart, G. *Weed Res.* **1997**, *37*, 171-179.
22) Houot, S.; Topp, E.; Yassir, A.; Soulas, G. *Soil Biol. Biochem.* **2000**, *32*, 615-625.
23) Yanze Kontchou, C.; Gschwind, N. *Appl. Environ. Microbiol.* **1994**, *60*, 4297-4302.
24) Mandelbaum, R.T.; Allan, D.L.; Wackett, L.P. *Appl. Environ. Microbiol.* **1995**, *61*, 1451-1457.
25) Radosevich, M.; Traina, S.J.; Hao, Y.L.; Tuovinen, O.H. *Appl. Environ. Microbiol.* **1995**, *61*, 297-302.
26) Bouquard, C.; Ouazzani, J.; Promé, J.C.; Briand, Y.M.; Plésiat, P. *Appl. Environ. Microbiol* **1997**, *63*, 862-866.
27) Struthers, J.K.; Jayachandran, K.; Moorman, T.B. *Appl. Environ. Microbiol.* **1998**, *64*, 3368-3375.
28) Topp, E.; Zhu, H.; Nour, S.M.; Houot, S.; Lewis, M.; Cuppels, D. *Appl. Environ. Microbiol.* **2000**, *66*, 2773-2782.
29) Topp, E.; Mulbry, W.M.; Zhu, H.; Nour, S.M.; Cuppels, D. *Appl. Environ. Microbiol.* **2000**, *66*, 3134-3141.
30) de Souza, M.L.; Wackett, L.P.; Bound-Mills, K.L.; Mandelbaum, R.T.; Sadowski, M.J. *Appl. Environ. Microbiol.* **1995**, *61*, 3373-3378.
31) de Souza, M.L.; Seffernick, J.; Martinez, B.; Sadowsky, M.J.; Wackett, L.P. *J. Bacteriol.* **1998**, *180*, 1951-1954.

32) Boundy-Mills, K.L.; Desouza, M.L.; Mandelbaum, R.T.; Wackett, L.P.; Sadowsky, M.J. *Appl. Environ. Microbiol.* **1997**, *63*, 916-923.
33) Sadowsky, M.J.; Tong, Z.K.; De Souza, M.; Wackett, L.P. *J. Bacteriol.* **1998**, *180*, 152-158.
34) Cook, A.M.; Beilstein, P.; Grossenbacher, H.; Hutter, R. *Biochem. J.* **1985**, *231*, 25-30.
35) Martinez, B.; Tomkins, J.; Wackett, L.P.; Wing, R.; Sadowsky, M.J. *J. Bacteriol.* **2001**, *183*.
36) Karns, J.S.; Eaton, R.W. *J.Agric. Food Chem.* **1997**, *45*, 1017-1022.
37) de Souza, M.L.; Newcombe, D.; Alvey, S.; Crowley, D.E.; Hay, A.; Sadowsky, M.J.; Wackett, L.P. *Appl. Environ. Microbiol.* **1998**, *64*, 178-184.
38) Radosevich, M.T., S.J.; Hao, Y.L.; Tuovinen, O.H. *Appl. Environ. Microbiol.* **1995**, *61*, 297-302.
39) Radosevich, M.; Traina, S.J.; Tuovinen, O.H. *J. Environ. Sci. Health Part B - Pesticides Food Contaminants and Agricultural Wastes* **1995**, *30*, 457-471.
40) Mulbry, W.W.; Zhu, H.; Nour, S.M.; Topp, E. *FEMS Microbiol. Lett.* **2002**, *206*, 75-79.
41) Rousseaux, S.; Hartmann, A.; Soulas, G. *FEMS Microbiol. Ecol.* **2001**, *36*, 211-222.
42) Rousseaux, S.; Soulas, G.; Hartmann, A. *FEMS Microbiol. Ecol.* **2002**, *41*, 69-75.
43) Topp, E. *Biol. Fertil. Soils* **2001**, *33*, 529-534.
44) Alvey, S.; Crowley, D.E. *Environ. Sci. Technol.* **1996**, *30*, 1596-1603.
45) Newcombe, D.A.; Crowley, D.E. *Appl. Microbiol. Biotechnol.* **1999**, *51*, 877-882.
46) Strong, L.C.; Mc Tavish, H.; Sadowsky, M.J.; Wackett, L.P. *Environ. Microbiol.* **2000**, *2*, 91-98.
47) Shapir, N.; Mandelbaum, R.T. *J. Agric. Food Chem.* **1997**, *45*, 4481-4486.
48) Cook, A.M.; Grossenbacker, H.; Hutter, R. *Experientia* **1983**, 39, 1191-1198.
49) Crawford, J.J.; Sims, G.K.; Mulvaney, R.L.; Radosevich, M. *Appl. Microbiol. Biotechnol.* **1998**, *49*, 618-623.
50) Crawford, J.J.; Traina, S.J.; Tuovinen, O.H. *Soil Sci. Soc. Am. J.* **2000**, *64*, 624-634.
51) Stucki, G.; Yu, C.W.; Baumgartner, T.; Gonzalez Valero, J.F. *Water Res.* **1995**, *29*, 291-296.
52) Ralebitso, T.K.; Senior, E.; van Verseveld, H.W. *Biodegradation.* **2002**, *13*, 11-19.
53) Jayachandran, K.; Stolpe, N.B.; Moorman, T.B.; Shea, P.J. *Soil Biol Bioch.* **1998**, *30*, 523-529.

54) Barriuso, E.; Houot, S.; *Soil Biol. Bioch.* **1996**, *28*, 1341-1348.
55) Ostrofsky, E.B.; Traina, S.J.; Tuovinen, O.H. *Soil Biol. Bioch.* **2002**, *34*, 1449-1459.
56) Lee, N.; Nielsen, P.H.; Andreasen, K.H.; Juretschko, S.; Nielsen, J.L.; Schleifer, K.-H.; Wagner, M. *Appl. Envrion. Microbiol.* **1999**, *65*, 1289-1297.
57) Gray, N.D.; Head, I.M. *Env. Microbiol.* **2001**, *3*, 481-492.
58) Hill, G.T.; Mitkowski, N.A.; Aldrich-Wolfe, L.; Emele, L.R.; Jurkonie, D.D.; Ficke, A.; Maldonado-Ramirez, S.; Lynch, S.T.; Nelson, E.B. *Appl. Soil Ecol.* **2000**, *15*, 25-36.
59) Martin-Laurent, F.; Philippot, L.; Hallet, S.; Chaussod, R.; Germon, J.C.; Soulas, G.; Catroux, G. *Appl. Environ. Microbiol.* **2001**, *67*, 2354-2359.
60) Miller, D.N.; Bryant, J.E.; Madsen, E.L.; Ghiorse, W.C. *Appl. Environ. Microbiol.* **1999**, *65*, 4715-4724.
61) Tsai, Y.-L.; Olson, B.H. *Appl. Environ. Microbiol.* **1991**, *57*, 1070-1074.
62) Shapir, N.; Goux, S.; Mandelbaum, R.T.; Pussemier, L.; *Can. J. Microbiol.* **2000**, *46*, 425-432.
63) Clausen, G.B.; Larsen, L.; Johnsen, K.; Radnoti de Lipthay, J.; Aamand, J.; *FEMS Microbiol. Ecol.* **2002**, *41*, 221-229.
64) Martin-Laurent, F.; Piutti, S.; Hallet, S.; Wagschal, I.; Philippot, L.; Catroux, G.; Soulas, G. *Pest Manag. Sci.* **2002**, In press.
65) Piutti, S.; Hallet, S.; Rousseaux, S.; Philippot, L.; Soulas, G.; Martin-Laurent, F. *Biol. Fert. Soils.* **2002**, In press.
66) Piutti, S.; Martin-Laurent, F.; Hartmann, A.; Lichtfouse, E.; Topp, E.; Soulas, G.; FEMS Microbiol. Lett. **2002**, In press
67) Wilson, M.S.; Bakermans, C.; Madsen, E.L. *Appl. Environ. Microbiol.* **1999**, *65*, 80-87.
68) Martin-Laurent, F.; Hallet, S.; Philippot, L.; Catroux, G.; Soulas, G. **2001** in 9th International Symposium on Microbial Ecology, pp. 355, International Society for Microbial Ecology, Amsterdam.Mandelbaum, R.T.; Wackett, L.P.; Allan, D.L. *Environ. Sci. Technol.* **1993**, *27*, 1943-1946.

Chapter 12

Detoxification of Pesticide Residues in Soil Using Phytoremediation

J. B. Belden, B. W. Clark, T. A. Phillips, K. L. Henderson, E. L. Arthur, and J. R. Coats

Department of Entomology, Iowa State University, Ames, IA 50011

During the past few years, we have conducted a series of experiments to investigate the potential of using plants as tools for the remediation of pesticide-contaminated soil. We have demonstrated that a blend of prairie grasses increases dissipation rates of several pesticides including metolachlor, trifluralin, and pendimethalin. However, in other studies, mulberry trees were not shown to influence pesticide dissipation. Additional studies have demonstrated that metolachlor movement in the soil column may be reduced by the presence of prairie grasses, bioavailability of dinitroaniline herbicides may be reduced during phytoremediation, and soil and leachate from remediated soil may have less toxicity than expected. Current studies within our laboratory are being conducted to determine the role of prairie grass blends in the phytoremediation procedure as compared to individual species and the role of plant uptake of pesticides in the phytoremediation process.

Introduction

Phytoremediation – the use of plants as a remediation agent for contaminated water or soil – has recently become widely investigated as a possible solution for many pollution problems. For example, plants have been used to remove heavy metals from soil, grasses have been used to remediate petroleum hydrocarbons, industrial solvents, and explosives from contaminated soil, and trees have been used to remove atrazine and industrial solvents from groundwater plumes (1). Phytoremediation is able to remediate such diverse contaminants due to the variety of mechanisms plants may use to either remove or detoxify contaminants. Heavy metals and some organic compounds may be removed by plant uptake such as the case for removal of heavy metals from soil and removal of volatile-organic compounds from groundwater. Organic compounds may be further degraded in the plant, while heavy metal uptake requires removal of the plant from the site. Plants are also capable of increasing degradation of organic compounds in the rhizosphere (root zone of the plant). This is often due to the plant releasing exudates from their roots, resulting in increased microbial activity (1); however, a few investigators have reported direct release of degrading enzymes, capable of biotransformation of organic compounds (2).

Our current research has focused on potential for using phytoremediation for cleanup of point-source pesticide contamination. During the manufacturing of pesticides, distribution to agrochemical dealerships, mixing of formulations, and loading of pesticides into tanks for application, there is a great potential for the occurrence of pesticide spillage. In fact, one study estimated that 90% of agrochemical dealerships in Iowa have pesticide contaminated soil and 50% of these sites will need remediation (3). High concentrations of pesticide contamination in soil can impact the environment in several ways, including leaching into groundwater, running off into nearby surface water, or directly impacting local soil organisms.

We have mostly concentrated on a set of pesticides – atrazine, metolachlor, pendimethalin, and trifluralin – that have been among the most heavily used pesticides in the corn-belt region of the United States for many years (4). Atrazine and metolachlor are moderately persistent in the environment and are relatively soluble in water. Studies have shown that they are two of the most common contaminants of ground and surface water (5). Pendimethalin and trifluralin are not very mobile in soil (6); however, they are persistent and tend to bioaccumulate (7).

In a series of studies, we have investigated the potential use of individual prairie grasses, a prairie grass mixture, and mulberry trees for the phytoremediation of these pesticides. Our initial experiments have been conducted to demonstrate that plants can survive in soil moderately contaminated with pesticides, and their presence increases the degradation rate of pesticides. Further experiments have been designed to evaluate the remediation system by examining pesticide movement in the soil column during remediation and evaluating bioavailability of the remaining residues. The latest experiments have been focused on the mechanisms involved with our phytoremediation strategy, including the impact of mixed grass species as compared to individual species and the amount of pesticide that is taken up by the plant versus degraded in the soil. The purpose of this chapter is to review our recent and current phytoremediation investigations. Full detail of the experiments is being published elsewhere.

Evaluation of Plants for Phytoremediation Potential

The prairie grasses we are investigating – yellow indiangrass, switchgrass, and big bluestem – are deep-rooted perennial plants with long growing seasons, they are tolerant to moderate levels of pesticides, and are commonly available as seed. Prairie grasses have also been shown to have phytoremediation potential for other organics such as petroleum hydrocarbons (8). We have chosen to use a mixture of the three grasses in most of our studies, because mixtures are used for prairie restoration and may have a wider range of degradative capacities in their rhizosphere.

Prairie Grasses - Microplot Study

In a four-year study, we investigated the effect of prairie grasses on the rate of pesticide dissipation in soil obtained from a contaminated agrochemical dealership site. The soil (loamy sand, 1.6% organic matter), originally containing 110 mg/kg pendimethalin and 10 mg/kg metolachlor, was treated with 25 mg/kg atrazine, trifluralin, and metolachlor. Microplots were constructed in plastic tubs (24 x 30-cm base and 18 cm-depth), which were kept in outdoor plots in Ames, IA during the summer and in the greenhouse during winter months. After an initial aging period of 30 days, individual microplots were either planted with a mixture of the three prairie grasses, or left unvegetated (n=4). Each microplot was sampled by taking three soil cores at various points in time up to 1,026 days. After 1,026 days, soil from each plot was allowed to dry and then mixed thoroughly. Soil samples were extracted with ethyl acetate and the concentration of pesticides was determined by gas chromatography with thermionic specific detection (9).

Initial measurements taken over the first 200 days of this study indicated a trend of increased atrazine and metolachlor dissipation in the prairie grass plots.

However, by the end of the study, less than 2% of the atrazine and less than 15% of initial metolachlor was recoverable. At this point, there was no significant difference between unvegetated and vegetated treatment of soil for either compound. The dinitroanaline herbicides, pendimethalin and trifluralin, were much more persistent. As shown in Figure 1, greater than 40%, and up to 70%, of the residues remained after 1,000 days. Lower amounts of dinitroaniline herbicides (pendimethalin and trifluralin) were recoverable from soil vegetated with prairie grasses (p=0.004). Individually statistical analysis did not show vegetation differences for pendimethalin, however the percentage of trifluralin residue remaining was significantly lower in vegetated columns (t-test, p<0.05).

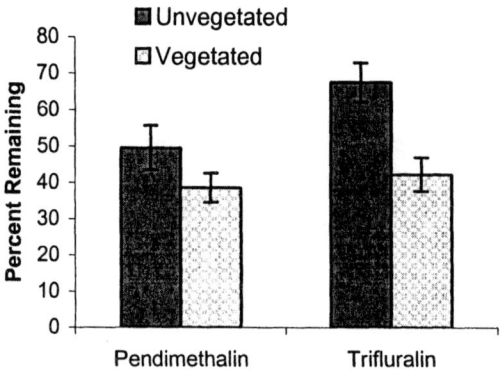

Figure 1. Vegetation with prairie grasses significantly decreased the percentage of pendimethalin and trifluralin remaining in soil after 1,000 days of remediation (F = 12.6, p = 0.004).

Prairie Grasses - Column Study

Further investigation of the prairie grass mixture for phytoremediation of pesticides was conducted using artificial soil columns. Eight soil columns were constructed in PVC (polyvinylchloride) pipe (10-cm diameter and 21-cm depth) with the bottom of the column secured with glass wool and aluminum screen. The base of the column was packed with 7 cm unfortified soil from an agronomic site that has not received pesticide application for over 15 years (sandy loam, 2.4% organic matter). An additional 14 cm of the same soil fortified with atrazine, alachlor, metolachlor, and pendimethalin at 25 mg/kg was added to the top of the column. The bulk density of the soil in the columns was 1.2 g/cm^3. After aging the columns for 60 days, half of the columns were planted with plugs of the prairie grass mixture previously described, while the remaining columns were left unvegetated.

The columns were placed in a greenhouse for 240 days and watered as needed to keep the columns moist, but not to cause water to come out of the bottom of the column. At the end of the study, a "storm event" was performed which entailed adding 608 ml water to each column (corresponding to 7.5 cm of rain). This amount of water resulted in leaching through the column. The leachate was extracted by solid-phase extraction. After 10 days without water, the columns were divided into three soil profile regions, and the soil was extracted by shaking with ethyl acetate. Extracts were analyzed by gas chromatography with thermionic specific detection. Full methods for the analysis techniques have been previously reported (9, 10).

Alachlor and atrazine degraded rapidly in the columns; less than 2% of the applied amount was recovered from the system and no differences were found between vegetated and unvegetated treatments. As shown in Figure 2, vegetation with prairie grasses did significantly reduce the total amount of metolachlor ($p<0.01$) and pendimethalin ($p<0.01$) recoverable from the soil column and leachate.

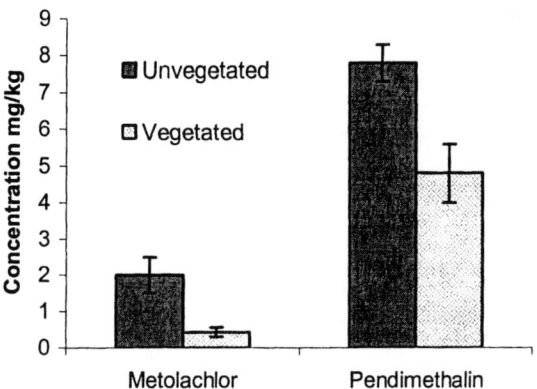

Figure 2. After 250 days of remediation, artificial soil columns planted with prairie grasses had significantly less extractable metolachlor and pendimethalin as compared with unvegetated soil columns.

Mulberry Trees

Mulberry trees (*Morus rubra*) have been suggested as potential phytoremediation tools due to root structure and composition of root exudates (11). This rapid-growth species of tree thrives in the Midwestern U.S.; therefore, we designed a study to investigate their use as a phytoremediation tool for pesticides. Soil fortified with 100 mg/kg atrazine, and 25 mg/kg trifluralin and metolachlor, was packed inside PVC pipe 15 cm in diameter and 30 cm long. Columns were allowed to age for 60 days to better reflect the type of pesticide residues found at agrochemical dealerships. Subsequently, ten artificial soil columns were planted with mulberry trees, and ten were left unvegetated. After 170 days, soil from 5 vegetated and 5 unvegetated columns was measured for pesticide concentration. Measurements were taken from the remaining columns at 330 days. Soil concentrations were determined using solvent extraction of the soil followed by analysis by gas chromatography and thermionic specific detection as previously described (9).

The trees grew at a slower rate than expected. Examination of the roots revealed limited growth in a twisted pattern, leading us to believe inhibition of root growth occurred, likely as a result of trifluralin. As shown in Figure 3, the presence of mulberry trees did not reduce the amount of pesticide present as compared to unvegetated controls. In fact, the mulberry containing soil columns contained significantly higher levels of metolachlor than did unvegetated columns ($p<0.01$). Several factors may influence the potential of mulberry trees for phytoremediation. First, as with all phytoremediation, damage to the plant by the contaminant may reduce the impact of the plant. Second, mulberry trees have been reported to release exudates in seasonal cycles with a greater release of phenols in the fall during senescing (11). However, because this experiment was conducted in a greenhouse, environmental factors may not have been appropriate to result in high releases of exudates. Third, the release of phenolic compounds in root exudate is likely to cause a shift in microbial populations (11). As a result, this shift may increase or decrease the degradation of contaminants on a contaminant-specific basis. Finally, soil obtained for this study had a previous history of metolachlor treatment. Therefore, if a population of microbes with metolachlor-degrading capabilities was already present, a decrease in this microbial activity in the mulberry columns may account for the results.

Evaluation of Phytoremediation Success Using Alternative Endpoints

Phytoremediation, as with all bioremediation, may take an extended period of time before remediation is successful. During this time, movement of the pesticide or metabolites generated in the process into biota, surface water, or

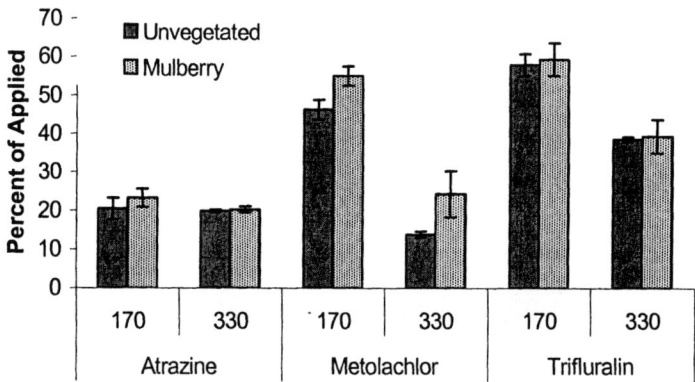

Figure 3. Percentage of applied atrazine, metolachlor, and pendimethalin recovered from columns as a whole at day 170 and 330. Presence of mulberry trees did not affect the percentage of atrazine and trifluralin remaining; however, mulberry did significantly increase the percentage of remaining metolachlor ($p<0.01$).

ground water may cause undesirable environmental effects. In order to evaluate our phytoremediation method thoroughly, we have evaluated many of our phytoremediation studies using alternative endpoints, in addition to the traditional approach of chemically measuring the concentration of the contaminants of interest.

Pesticide Movement within the Soil Column During Phytoremediation

In the study previously described as the "Column Study", concentrations of pesticide were measured throughout the column and in leachate recovered from the bottom of the column after a "storm" event. Figure 4 illustrates the percentage of total recoverable metolachlor obtained from the leachate. Interestingly, the vegetated columns not only reduced the total amount of metolachlor present as previously noted (see Figure 2), but also decreased downward leaching of metolachlor. After 160 days of remediation, vegetation resulted in a two-fold decrease in the total amount of metolachlor recovered from the system, while a five-fold decrease was noted for the amount of metolachlor that was recovered from leachate. After 250 days, the presence of vegetation resulted in a four-fold decrease recorded for total metolachlor, while a 20-fold decrease was noted for leachate.

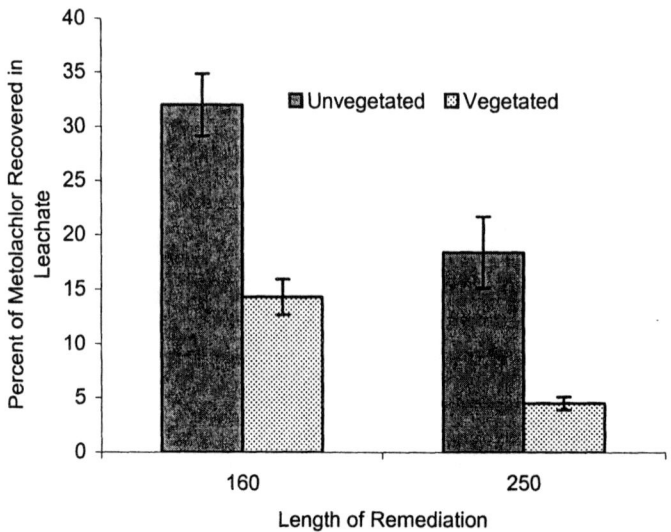

Figure 4. Percentage of total metolachlor recovered from the system, that was recovered in leachate. Significantly less metolachlor was recovered in leachate from vegetated columns as compared with unvegetated columns (p<0.01).

Bioavailability of Pesticide Residues

As compounds age in soil, they often have reduced bioavailability, potentially resulting in decreased environmental risk (12); however, the rate of bioremediation may also decrease. During phytoremediation, plant-induced changes in the soil environment may potentially change bioavailability as well; therefore it is important to monitor bioavailability while evaluating phytoremediation techniques. We have used two main assays for determining bioavailability in remediating soil. The first assay, an 8-day earthworm bioavailability test, was conducted as previously described (12). Earthworms *(Eisenia fetida)* were exposed to the contaminated soil, followed by analysis of the worms and test soil for pendimethalin and metolachlor. The ratio of the concentration in the worm compared to the concentration in the soil was determined (biological accumulation factor; BAF). The bioassay was conducted on soil immediately after fortification of pendimethalin and with soil obtained during the phytoremediation experiment previously described as the Column Study. The ratio of the BAF determined for the remediated soil was divided by the BAF determined for fresh residues to obtain percent bioavailability for earthworms.

The second bioassay used to evaluate bioavailability was lettuce seedling growth. Percentage inhibition was determined seven days after adding seeds to soil. Multiple concentrations of pendimethalin and metolachlor were evaluated as fresh residues to obtain dose-response relationships. Seedling growth was then measured in soil obtained from the previously described Column Study. Comparison of pesticide concentrations in the Column Study soil to the dose-response data obtained for each pesticide suggested that pendimethalin was the major toxicant in the system to lettuce. Therefore, the seedling inhibition rates determined for Column Study soil were used to calculated effective soil concentrations using the dose-response curve for pendimethalin. The effective concentration was then divided by the measured soil concentration of pendimethalin to obtain the percent bioavailability for lettuce.

As shown in Figure 5, the bioavailability of pendimethalin, as measured by the earthworm assay and the lettuce assay, was reduced by the presence of vegetation and by section after 160 days. The section effect is likely due to the addition of organic matter (in the form of potting soil) into the top section during the addition of plants (and potting soils plugs into controls). Vegetation also adds organic matter to soil and may increase microbial activity and the turn-over rate of organic matter. These processes could be responsible for the vegetation related decrease in bioavailability.

The decrease in bioavailability was of greater magnitude for lettuce as compared to earthworm uptake. The difference is likely due to the uptake mechanism differences between the species (plant and animal). Lettuce uptake is primarily through partitioning of the toxicant from soil, into soil water, and then into the plant root. Since earthworms ingest soil, enzymes and other gut factors may aid in the uptake of the pesticide. Chemical differences between lettuce and earthworm cuticles may also be very important. When calculating bioavailability, it is important to acknowledge that large differences may exist between types of biota, and between biological and chemical endpoints.

Current Research in Understanding and Improving Phytoremediation Techniques

In order to improve and fully utilize phytoremediation for cleaning up pesticide-contaminated soil, we need more basic knowledge about the process. We are actively conducting research in two areas that should improve our understanding of the role of plants within the system. In the first area, we are evaluating the fate of radiolabeled pesticides within prairie grass-soil systems. Plants may increase dissipation rate through uptake of the pesticide or through increased degradation in the rhizosphere. Understanding the role of plants is

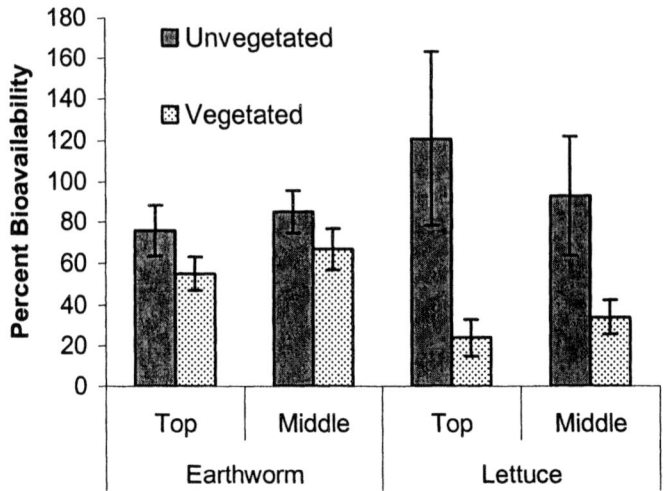

Figure 5. Bioavailability of pendimethalin as measured by earthworm uptake and lettuce seedling growth was reduced by the presence of vegetation ($p<0.05$, $p<0.01$, respectively) in the top and middle sections of soil columns.

crucial in efforts to improve phytoremediation technologies. Currently, separate studies investigating the fate of atrazine, metolachlor, and pendimethalin are being conducted.

The second area of study involves evaluating the role of species type and species mixture on phytoremediation capability. In one experiment, we investigated the effect of prairie grasses, individually and as a mixture, on the dissipation rate of pendimethalin in a sandy loam soil (2.7% organic matter) collected from a corn field near Ames, IA that has a history of no pesticide treatment. The soil was fortified with 25 mg/kg pendimethalin and aged for 30 days. After aging, 500 g of treated soil was added to each of 20 cones (6.5-cm diameter and 25.4-cm depth), and plugs were added: potting soil only (control), big bluestem, yellow indiangrass, switchgrass, or a mixture of all three prairie grasses (4 reps/treatment). After 180 days of remediation, soil from each cone was mixed thoroughly, extracted with ethyl acetate, and analyzed by gas chromatography (9).

After 180 days, significantly lower amounts of pendimethalin were recovered in soil vegetated with switchgrass, big bluestem, and a mixture of all three prairie grasses compared to unvegetated soil ($p<0.05$; Figure 6). These results indicate that switchgrass and big bluestem are more effective at increasing dissipation rates of pendimethalin as compared to yellow indiangrass. The mixture of prairie grasses had a similar effect on dissipation of pendimethalin as big bluestem and switchgrass. Switchgrass had the least

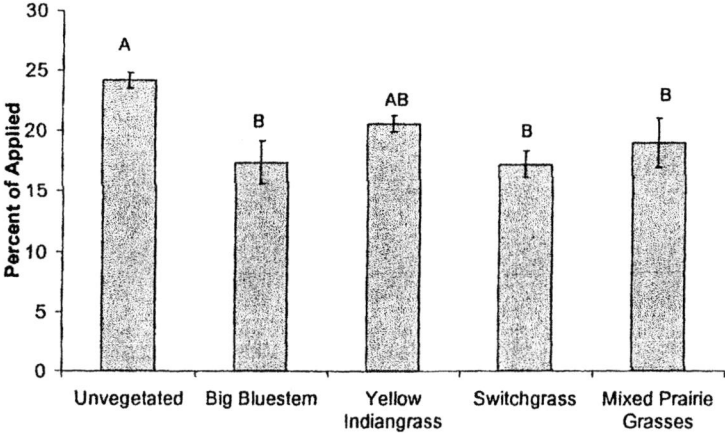

Figure 6. Big bluestem, switchgrass, and mixed prairie grasses all significantly reduced the amount of recovered pendimethalin after 180 days of remediation ($p<0.05$). Treatments with the same letter are not significantly different.

biomass at the end of the study, indicating that the results are not strictly tied to growth and productivity of the grasses.

Conclusions

We have found substantial evidence that the presence of vegetation can increase dissipation rates of pesticide residues in soil. Further evidence implies that vegetation may stabilize the pesticide residues, decreasing the potential for leaching and uptake into biota; thus, phytoremediation may prove to be a valuable tool in clean up of moderately contaminated sites. However, several considerations should be made regarding this technology:

1]. Plant selection is crucial. In the studies presented here, some grass species increased dissipation, one grass species had little observable effect, and mulberry trees may have inhibited metolachlor dissipation. Additionally, as seen in the mulberry experiment, pesticide damage to the plant may reduce plant growth and therefore, the impact of the plant on the environment.

2]. The process is slow. Our results have demonstrated that plants may help stabilize the system, however, if severe environmental impact is imminent, other remediation may be necessary.

3] Changes in bioavailability may hinder the complete clean-up of the site. Potential exists for aged residues, which are not available to degrading organisms to remain on the site. If strict concentration-based action limits exist for the site, the remediation technique may not be successful. However, if the compounds are not available for degradation, it is likely that the compound has limited availability to sensitive organisms or potential to leach into groundwater.

Further research studying the mechanisms and factors influencing phytoremediation may help us to improve the technology by increasing the rate of pesticide degradation and increasing our are capabilities in stabilizing remaining residues.

Acknowledgements

Partial financial support for this project was provided by the Center for Health Effects of Environmental Contaminants (CHEEC) at the University of Iowa. This journal paper of the Iowa Agriculture and Home Economics Experiment Station, Ames, Iowa, Project No. 3187, was supported by Hatch Act and State of Iowa funds.

References

1. United States Environmental Protection Agency. 2000. Introduction to phytoremediation. U.S. EPA Office of Research and Development, Washington D.C., EPA/600/R-99/107.
2. Siciliano, S.D., H. Goldie, and J.J. Germida. 1998. Enzymatic activity in root exudates of Dahurian wild rye (*Elymus dauricus*) that degrades 2-chlorobenzoic acid. *J. Agric. Food Chem.* 46:5-7.
3. Gannon, E. 1992. Environmental Clean-up of Fertilizer and Agrichemical Dealer Sites - 28 Iowa Case Studies. Iowa Natural Heritage Foundation, Des Moines, IA. pp. 5-19.
4. National Agricultural Statistics Service. 1997. Agricultural Chemical Usage, 1996. U.S. Department of Agriculture, Washington, D.C.

5. Gilliom, R.J., J.E. Barbash, D.W. Kolpin, and S.J. Larson. 1999. Testing water quality for pesticide pollution. *Environ. Sci. Technol.* 33: 164A-169A.
6. Zheng, S.Q., J.F. Cooper, and P. Fontanel. 1993. Movement of pendimethalin in soil of the south of France. *Bull. Environ. Contam. Toxicol.* 50:492-498.
7. United States Environmental Protection Agency. 1999. Persistent bioaccumulative toxic (PBT) chemicals: final rule. 40 CFR Part 372.
8. Aprill W., and R.C. Sims. 1990. Evaluation of the use of prairie grasses for stimulating polycyclic aromatic hydrocarbon treatment in soil. *Chemosphere* 20:253-265.
9. Anderson, T.A., E.L. Kruger, and J.R. Coats. 1994. Enhanced degradation of a mixture of three pesticides in the rhizosphere of a pesticide-tolerant plant. *Chemosphere* 28:1551-1557.
10. Belden, J.B., and M.J Lydy. 2000. Analysis of multiple pesticides in urban storm water using solid-phase extraction. *Arch. Environ. Contam. Toxicol.* 38: 7-10.
11. Hegde, R.S., and J.S. Fletcher. 1996. Influence of plant growth stage and season on the release of root phenolics by mulberry as related to development of phytoremediation technology. *Chemosphere* 32:2471-2479.
12. Kelsey, J.W., and M. Alexander. 1997. Declining bioavailability and inappropriate estimation of risk of persistent compounds. *Environ. Toxicol. Chem.* 16:582-585.

Chapter 13

Remediation of Halogenated Fumigant Compounds in the Root Zone by Subsurface Application of Ammonium Thiosulfate

Sharon K. Papiernik[1], Frederick F. Ernst[2], Robert S. Dungan[1], Wei Zheng[2], Mingxin Guo[2], and Scott R. Yates[1]

[1]George E. Brown Jr. Salinity Laboratory, Agricultural Research Service, U.S. Department of Agriculture, 450 West Big Springs Road, Riverside, CA 92507–4617
[2]Department of Environmental Science, University of California, Riverside, CA 92521

Fumigants are used for control of nematodes, fungi, weeds, and insects in high-cash-value crops. Because of their high volatility, a large fraction of the applied mass may be volatilized from the soil surface following application. Emission reduction strategies are needed to prevent adverse human or environmental health impacts. Some emission-reduction strategies, such as tarping the soil surface with impermeable plastic, increase containment of fumigants in the soil. The resulting fumigant residues could cause atmospheric contamination once the tarp is disrupted, groundwater contamination if leaching is allowed, or phytotoxicity to the crop planted following fumigation. Previous research has shown that thiosulfate compounds, including ammonium thiosulfate (ATS), abiotically react with and detoxify halogenated fumigants in soil and water. In these experiments, we investigated subsurface application of ATS to reduce fumigant concentrations in the root zone, preventing off-site transport following fumigation. Results indicated that halogenated fumigants were dissipated more rapidly in soil receiving ATS application compared to those receiving water only. First-order dissipation half-lives were ≤1 day for the halogenated compounds 1,3-dichloropropene and propargyl bromide. For methyl isothiocyanate, a non-halogenated fumigant that does not undergo reaction with ATS, application of ATS had no impact on the rate of dissipation in soil. These results suggest that subsurface application of ATS may be useful for the root-zone remediation of halogenated compounds.

© 2004 American Chemical Society

Fumigants are used worldwide for the control of soil-borne pests, such as nematodes, fungi, weeds, and insects in high-cash-value crops. In the United States, fumigants are intensively used for strawberry and tomato production in California and Florida. The popular fumigant methyl bromide (MeBr) is being phased out in the United States and other signatories to the Montreal Protocol, and the phase-out is to be complete by 2005. Other chemical alternatives, particularly 1,3-dichloropropene (1,3-D), chloropicrin, and methyl isothiocyanate (MITC), are already available as partial replacements for MeBr. The potential of additional compounds, including iodomethane and propargyl bromide (PrBr), to serve as soil fumigants is being assessed.

Fumigant compounds have high vapor pressures and are transported in soils largely in the gas phase. Without adequate containment, a significant proportion of the applied mass may be volatilized from the soil surface following soil application. For example, laboratory and field research has indicated that 30-60% of the applied 1,3-D is lost following subsurface application to uncovered soil (*1, 2*).

Because fumigant compounds have broad toxicity and are highly mobile, reducing atmospheric emissions is vital to their continued use. The use of several fumigant compounds, including 1,3-D, has been restricted due to air quality concerns (*3, 4*). Fumigation methods to increase containment and thus prevent air and water contamination by fumigant compounds are being developed. Containment also maximizes efficacy by increasing the time for which high fumigant concentrations are present in the soil. Management practices to reduce emissions include tarping the soil surface with impermeable plastic, increasing the soil water content to reduce gas-phase diffusion, and increasing the rate of fumigant transformation in the soil. Tarping the soil surface with a continuous sheet of impermeable plastic can drastically increase containment, resulting in near-zero emissions and reducing the fumigant application rate required to achieve adequate pest control (*5*).

Another promising approach to reducing emissions involves enhancing fumigant degradation at the soil surface. Previous research has indicated that nucleophilic compounds such as ammonium thiosulfate (ATS) rapidly transform and detoxify halogenated fumigants such as MeBr, iodomethane, chloropicrin, PrBr, and 1,3-D in aqueous solution and in soil (*6, 7*). It is postulated that the reaction between ATS and these halogenated fumigants is via S_N2 nucleophilic substitution, with thiosulfate replacing the halide (*6*). Similar reactions have been observed for reaction between ATS and other halogenated pesticides (*8*). Application of ATS at the soil surface can significantly reduce emissions of halogenated fumigants by forming a reactive barrier at the soil surface. Surface application of ATS reduced cumulative emissions of 1,3-D by >50% (*7*), and MeBr emissions by ~90% (*9*). Addition of nucleophilic compounds has no impact on compounds which do not undergo nucleophilic substitution, such as the non-halogenated fumigant MITC. Thiosulfate compounds are commonly used as fertilizers, and integration of ATS application and fumigation with 1,3-D showed no phytotoxic symptoms in tomato planted following fumigation (*7*).

Emissions-reduction strategies are required to minimize air contamination by fumigants and to increase efficacy. However, retention of significant concentrations of fumigants in soil following the fumigation period may result in air and groundwater contamination and crop phytotoxicity. For example, a large flux of fumigants to the atmosphere is observed following the removal of low-permeability tarps (5). Groundwater contamination by fumigants has been observed when heavy precipitation occurs when fumigant concentrations are relatively high, i.e., shortly after fumigation in conventional applications (10). Phytotoxicity to the crop planted following fumigation is possible where fumigant residues remain (11).

In some current soil fumigation practices using 1,3-D, the fumigant is applied with water through subsurface drip irrigation lines. Similar application methods have been proposed for potential alternative fumigants such as PrBr. In these experiments, we investigated the potential for subsurface application of ATS to reduce soil concentrations of 1,3-D and PrBr following soil fumigation, thus reducing the threat of off-site transport and potential phytotoxic effects on crops.

Methods

Four concrete mesocosms (3 m long by 1.5 m wide by 1.6 m deep) were filled with washed river sand to a bulk density of 1.4 Mg m^{-3}. Beds were formed at the soil surface; beds were 50 cm across the top and 20 cm high, as indicated in Figure 1. As part of an emissions-reduction experiment (results not presented here), different management practices were represented in the mesocosms, including a bare soil surface, bed tarped with HDPE, or bed tarped with Hytibar® (a low-permeability film). Subsurface drip lines (HDPE) were installed 15 cm below the bed surface. Fumigants were mixed with 24 L of water in HDPE carboys, which were then connected to the drip lines and pressurized to apply the fumigant mixture to the mesocosms. Application required 2-3 hours. Fumigant application rates were typical of field application: 10 gal ac^{-1} of 1,3-D C-35 (In-Line®); 80 lb ac^{-1} of PrBr; and 17 gal ac^{-1} of Vapam® (MITC precursor). These application rates corresponded to 0.2 to 0.3 mol of 1,3-D, PrBr, and MITC added to each mesocosm. After a 10-day fumigation period, ATS was drip-applied to half of the mesocosms in the same manner (380 mL of Thio-Sul® fertilizer in 24 L water, ~2 mol of ATS). The remaining mesocosms received 24 L of tap water (no ATS) to indicate fumigant concentrations remaining in the root zone with no ATS treatment.

Soil gas samples were collected to determine fumigant concentrations remaining in the soil. Teflon tubing (1-mm ID) was buried during bed construction and tubes terminated from 20 to 80 cm below the soil surface throughout the bed cross-sectional area (locations indicated in Figure 1) to provide information on the distribution of fumigant compounds in the root zone. Gas samples (50 mL) were collected on activated charcoal adsorbent tubes; syringes were used to apply vacuum and to measure the gas volume sampled. This approach has been successfully used in previous experiments to monitor soil gas concentrations (12). Fumigant compounds were extracted from charcoal using 3 mL of acetone. Fumigants in acetone extracts were identified and quantified by gas chromatography, using an electron capture detector for PrBr and 1,3-D and a nitrogen-phosphorus detector for MITC. Soil gas concentrations of fumigants were

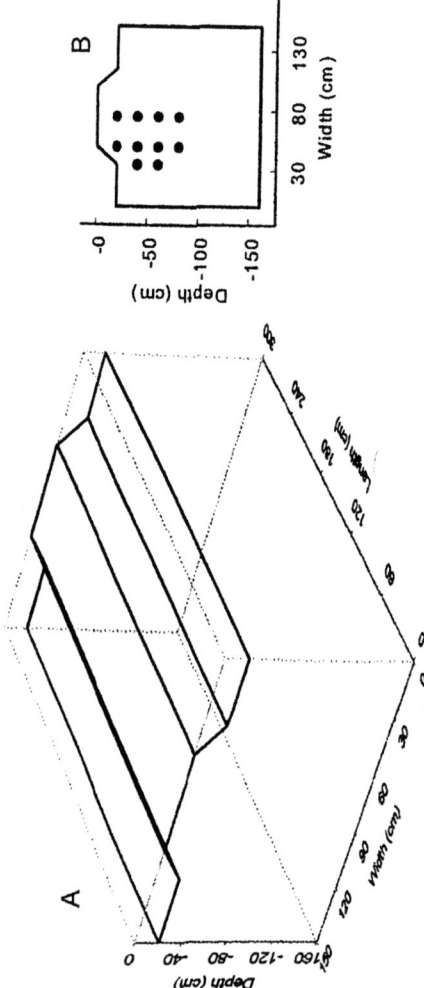

Figure 1. (A) Dimensions of mesocosms and (B) location of soil gas samplers in xy plane.

measured just prior to ATS application, and were monitored for 7 days following ATS application. Concentration data were kriged to construct contour maps of soil gas concentrations throughout the soil profile. The volume contained under the contours was determined to indicate the mass of fumigant remaining in the monitored zone of each mesocosm with time.

Results and Discussion

Initial soil gas concentrations varied over an order of magnitude due to the management practices used in the experiments. Residual concentrations were greater in mesocosms which had better containment. For all fumigant compounds, tarping the bed with an impermeable film resulted in the greatest containment (Figure 2). Tarping with a more permeable film (HDPE) resulted in less effective containment, and leaving the soil surface bare resulted in the lowest soil concentrations of fumigants following the 10-day fumigation period (Figure 2).

These soil concentrations indicated the potential for highly effective containment to result in higher post-fumigation concentrations in the root zone, which may result in more effective control with lower application rates, but may also result in air and groundwater contamination and crop phytotoxicity. Destroying the residual fumigant in the soil decreases the threat of environmental contamination by these compounds and diminishes the potential for crop injury upon planting.

Depletion of Fumigants in the Root Zone by ATS Application

Fumigant concentrations in soil gas samples were kriged to construct contours of the concentration of each compound at each sampling time (Figure 3). Because of the high variability in initial concentrations (Figure 2), the concentration at each sampling point was normalized to the concentration measured at that location just prior to ATS/water application in Figure 3. Results for each sampling time indicated that fumigant concentrations in the root zone were depleted to a greater extent in tanks treated with ATS than in those receiving water only.

Integrating the area under the concentration contours provided a means of estimating the mass remaining in each mesocosm at each time. For the example given in Figure 3, the total PrBr in the ATS-treated tank at 52 hours after ATS application (Figure 3A) was ~20% of the PrBr present prior to ATS application. In the mesocosm receiving no ATS (Figure 3B), the PrBr mass remaining at 52 hours was ~60% of the initial PrBr.

Subsurface application was effective at reducing the residual concentrations of halogenated fumigants (in this study, PrBr and 1,3-D), but had no impact on a non-halogenated fumigant (MITC). Each mesocosm was simultaneously treated with a solution containing 1,3-D, PrBr, and MITC. Within the first day following ATS application, the mass (normalized to initial conditions) of each halogenated

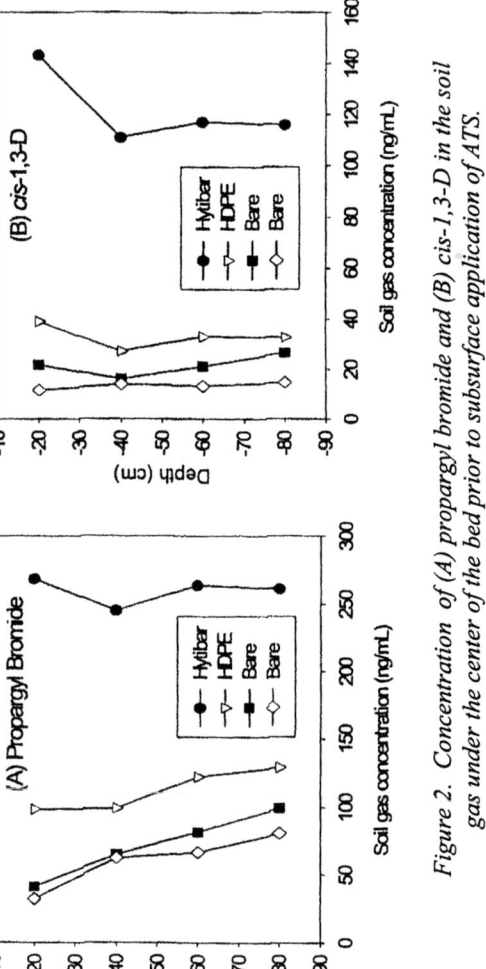

Figure 2. Concentration of (A) propargyl bromide and (B) cis-1,3-D in the soil gas under the center of the bed prior to subsurface application of ATS.

175

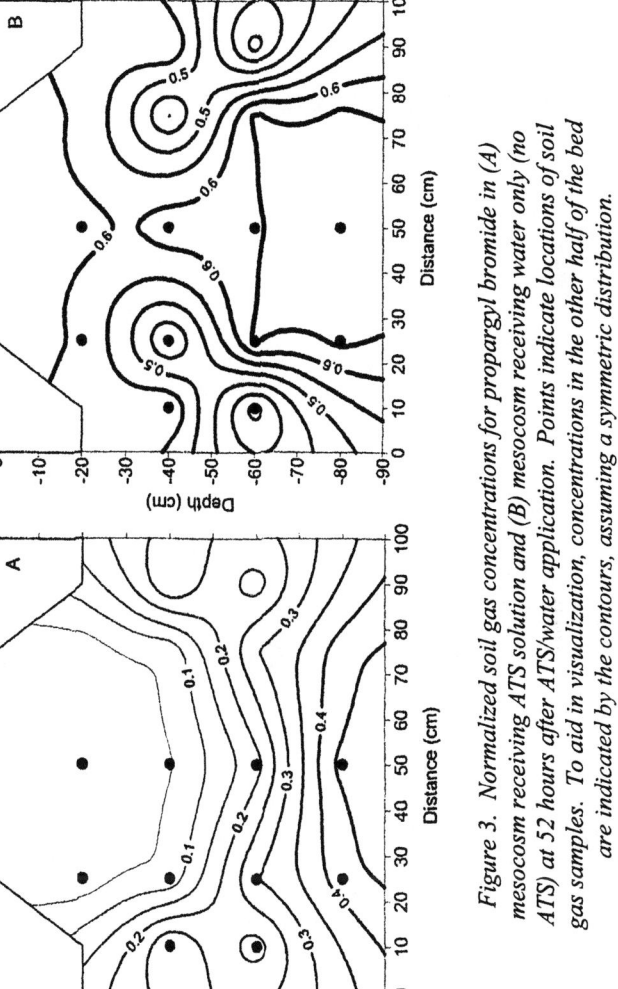

Figure 3. Normalized soil gas concentrations for propargyl bromide in (A) mesocosm receiving ATS solution and (B) mesocosm receiving water only (no ATS) at 52 hours after ATS/water application. Points indicate locations of soil gas samples. To aid in visualization, concentrations in the other half of the bed are indicated by the contours, assuming a symmetric distribution.

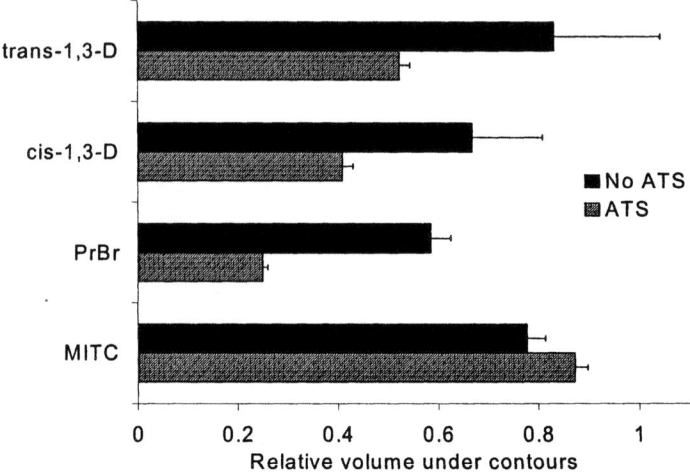

Figure 4. Mass remaining in replicate mesocosms at 28 days after ATS/water application relative to the mass present just prior to application. (Mass is indicated by the volume under the concentration contours.) Values indicate the mean of two mesocosms and error bars indicate the standard error.

compound was significantly lower than that in tanks receiving no ATS (Figure 4). The normalized mass of MITC was not significantly different in ATS-treated and non-treated tanks (Figure 4).

Fitting a first-order dissipation model ($C=C_0 e^{-kt}$) indicated that the decay constant (k, h^{-1}) for halogenated fumigants was approximately 2.5 to 5 times greater in ATS-treated tanks than in non-treated tanks (Figure 5, Table I). For PrBr, more than half of the residual post-fumigation PrBr was depleted within the first day following subsurface ATS application (Figure 4 and 5, Table I). In ATS-treated mesocosms, 1,3-D isomers were dissipated more slowly than PrBr, and the first-order half-life for 1,3-D was approximately twice that of PrBr (Table I). These results are consistent with laboratory measurements of ATS reaction with fumigants in soil, in which degradation of PrBr by ATS in soil occurred at approximately twice the rate of 1,3-D degradation (*6*). The slower dissipation of 1,3-D may be due to the slower rate of reaction between ATS and 1,3-D in solution and because of compound-specific differences in the air-soil-water partitioning (*6*). Because MITC is not affected by ATS application (no chemical reaction occurs between MITC and ATS), MITC was dissipated in ATS-treated and non-treated tanks at approximately the same rate (Figure 5).

Table I. Rate of Fumigant Dissipation

Fumigant	First-order dissipation constant (h^{-1}) ± standard error	
	ATS	No ATS
Propargyl bromide	0.046 ± 0.009 (0.96)	0.009 ± 0.002 (0.88)
1,3-Dichloropropene (*cis+trans*)	0.023 ± 0.004 (0.96)	0.009 ± 0.001 (0.96)
MITC	0.0060 ± 0.0005 (0.99)	0.0072 ± 0.0005 (0.99)

NOTE: Values in parentheses are r^2 for nonlinear regression.

Initial fumigant concentrations in the mesocosms varied by an order of magnitude (Figure 2), and each treatment (ATS and no ATS) included one tank that had a bare soil surface during fumigation (low post-fumigation concentrations). Results were consistent between replicate tanks (Figures 4 and 5), indicating that with proper distribution, subsurface application of ATS can be expected to consistently reduce post-fumigation concentrations of halogenated fumigant compounds.

These results indicate that for situations in which residual concentrations of halogenated chemicals subject to nucleophilic substitution are present in the soil, subsurface application of nucleophilic compounds such as ATS may be valuable for reducing the threat of leaching, runoff, volatilization, and phytotoxicity to crops. This approach may be particularly useful for remediation of chemicals applied via drip irrigation, including soil fumigants and other pesticides, since the application of ATS through existing drip lines would result in a minimal additional expense. Thiosulfate compounds, including ammonium, potassium, and calcium thiosulfate, are commonly used as fertilizers and pose little toxicity threat. Since ATS has been shown to degrade other halogenated agrochemicals, such as some chlorinated acetanilide herbicides (*8*), this approach has promise for root-zone remediation of many soil-applied chemicals.

Acknowledgments

We acknowledge the assistance of Qiaoping Zhang and Christian Taylor in obtaining the experimental data. In Line® was provided by Dow AgroSciences. This research was funded in part by the Methyl Bromide Transitions Program, award number 00-51102-9551.

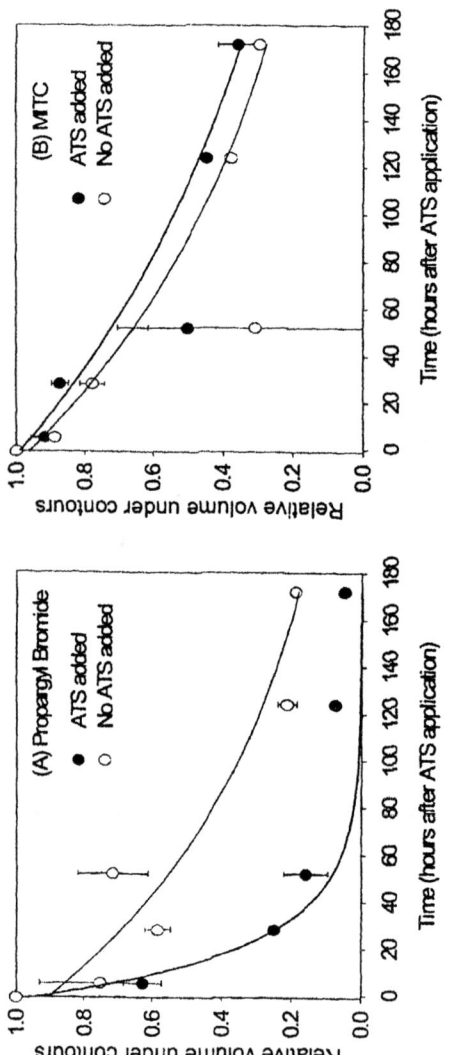

Figure 5. Decrease in (A) propargyl bromide and (B) MITC in the root zone with time. Values indicate the total volume contained under the concentration contours; the volume at each time was normalized to the volume observed just prior to injection of ATS/water. Values are the mean of two mesocosms and error bars represent the standard error. Lines indicate regression to a first-order kinetic model. For MITC (B), regression omitted spurious 52-hour data.

References

1. Wang, D.; Yates, S. R.; Ernst, F. F.; Knuteson, J. A. Volatilization of 1,3-dichloropropene under different application methods. *Water, Air, Soil Pollut.* **2001,** *127,* 109–123.
2. Gan, J.; Becker, J. O.; Ernst, F. F.; Hutchinson, C.; Knuteson, J. A.; Yates, S. R. Surface application of ammonium thiosulfate fertilizer to reduce volatilization of 1,3-dichloropropene from soil. *Pest Manag. Sci.* **2000,** *56,* 264–270.
3. California Department of Pesticide Regulation (CDPR). DPR approves limited use of soil fumigant. CDPR: Sacramento, CA, 1994; News Release 94-42.
4. California Department of Pesticide Regulation (CDPR). California Management Plan: 1,3-Dichloropropene. CDPR: Sacramento, CA, Jan. 30, 2002.
5. Wang, D.; Yates, S. R.; Ernst, F. F.; Gan, J.; Jury, W. A. Reducing methyl bromide emission with a high barrier plastic film and reduced dosage. *Environ. Sci. Technol.* **1997,** *31,* 3686–3691.
6. Wang, Q.; Gan, J.; Papiernik, S. K.; Yates, S. R. Transformation and detoxification of halogenated fumigants by ammonium thiosulfate. *Environ. Sci. Technol.* **2000,** *34,* 3717–3721.
7. Gan, J.; Yates, S. R.; Knuteson, J. A.; Becker, J. O. Transformation of ,13-dichloropropene in soil by thiosulfate fertilizers. *J. Environ. Qual.* **2000,** *29,* 1476–1481.
8. Gan, J.; Wang, Q.; Yates, S. R.; Koskinen, W. A.; Jury, W. A. Dechlorination of chloroacetanilide herbicides by thiosulfate salts. *Proc. Natl. Acad. Sci. U.S.A.* **2002,** *99,* 5189–5194.
9. Gan, J.; Yates, S. R.; Becker, J. O.; Wang, D. Surface amendment of fertilizer ammonium thiosulfate to reduce methyl bromide emission from soil. *Environ. Sci. Technol.* **1998,** *32,* 2438–2441.
10. Loria, R.; Eplee, R. E.; Baier, J. H.; Martin, T. M.; Moyer, D. D. Efficacy of sweep-shank fumigation with 1,3-dichloropropene against *Pratylenchus penetrans* and subsequent groundwater contamination. Plant Dis. **1986,** *70,* 42–45.
11. Noling, J. W. Nematode Management in Carrots; Report ENY-021; Entomology and Nematology Department, Florida Cooperative Extension Service, University of Florida: Lake Alfred, , March 1999.
12. Wang, D.; Yates, S. R.; Ernst, F. F.; Gan, J.; Gao, F.; Becker, J. O. Methyl bromide emission reduction with field management practices. *Environ. Sci. Technol.* **1997,** *31,* 3017–3022.

Chapter 14

Detoxification and Destruction of PCBs, CAHs, CFCs, and Halogenated Biocides in Soils, Sludges, and Other Matrices Using Na/NH$_3$

Charles U. Pittman, Jr.

Department of Chemistry, Mississippi State University, Mississippi State, MS 39762

Many toxic or hazardous organic compounds contain one or more chlorine atoms which are central to their biological effects. Examples include pesticides, such as lindane, **1**, and Mirex, **2**, polychlorinated biphenyls (PCBs), **3** and **4**, chlorinated aliphatic hydrocarbons (CAHs), **5**, tetrachloroethylene, **6**, dioxins, **7**, and chlorinated aromatic compounds such as 1,2-dichlorobenzene, **8**, or pentachlorophenol, **9**. Although not toxic, chlorofluorocarbons such as, **10**, are environmental hazards since they play an important role in distruction of ozone in the stratusphere. Since many of these compounds are widely dispersed in the environment, methods to remediate locations where they are pollutants and destroying them in cost efficient ways is an important goal. Processes to destroy chemical warfare agents, non-halogenated pesticides, polynuclear aromatic hydrocarbons, toxic metal compounds and germicides are also needed.

1   **2**   **3**

4 (2,2',6,4'-tetrachlorobiphenyl) **5** CH₃CCl₃ **6** C₂Cl₄

7 (tetrachlorodibenzodioxin) **8** (1,2-dichlorobenzene) **9** (chlorinated cyclohexanol) **10** CF₃Cl

Many methods to destroy these compounds are available including incineration,[1,2] wet air oxidation,[3] catalytic dehydrochlorination,[4] reaction with superoxide,[5] photolysis in the presence of hydrogen donors,[6] transition metal-promoted reductive dechlorinations using sodium borohydride or alkoxyborohydrides,[7,8] electrolytic reductions,[9] hydrogenolyses,[10] iron-pranoted dechlorination[11] and thermolysis over solid bases such as $CaO/Ca(OH)_2$.[12] All these methods have drawbacks, especially when the toxic pollutants are already distributed in the soil, sludges, ground or surface waters, etc.

When concentrated samples are available, even simple methods like combustion require special treatments to remove the HCl generated.[13] Incineration of PCBs and other chlorinated organic compounds can produce small amounts of highly toxic polychlorinated dibenzofurans and dibenzodioxins (e.g. **7**). Therefore, methods which mineralize the chlorine as simple chloride salts could have a distinct advantage. High temperature reductions with $NaBH_4$ will reduce PCBs.[14] Lower temperature sodium-based reductions in hydrocarbon media will generate NaCl and they were examined extensively.[15,16] Unfortuately the kinetics of reduction were slow. Direct thermolysis of PCBs over $CaO/Ca(OH)_2$ or $MgO/Mg(OH)_2$ will mineralize the chlorine but high temperatures are needed for efficiency[12] and, when applied in soils, migration of the PCBs may occur.

Alkali or alkaline metals dissolve readily in liquid ammonia to give blue solutions of the metal cation and solvated electrons.[30] Independently, our group[17-23] and Commodor Solution Technologies[24-29] reasoned that solvated electron solutions would rapidly dechlorinate chlorinated organics and this concept was then demonstrated using both calcium and sodium. These dechlorinations are very fast at room temperature and even –60 °C! Evidence was presented they often occur at diffusion controlled rates for chlorinated aromatic compounds and CAHs.[19] While it has long been known that dechlorination of organic compounds[30,31] will occur readily in solvated electron ammonia solutions,[30,31] the potential application of this reaction to remediation of wet soils or sludges was simply never considered. Instead, it was assumed solvated electrons (or Na) would react rapidly with water. Thus, it was thought that large amounts of sodium (or Ca etc) would be consumed, making such reductions impractical. This assumption has now been shown to be wrong.[17-25] Table 1 shows that PCB-contaminated soils can be readily remediated. Simply slurrying the soil in NH_3 followed by adding Na or Ca at ambient temperature results in excellent remediation. Furthermore, after adding Na or Ca, the reduction is completed in seconds or a few minutes. All the soils in Table 1 were wet. The last three examples contained more than 20% wt. water.

The amount of sodium required to reach the same degree of decontamination of polluted soils was frequently higher when Na was first dissolved in NH_3/ followed by slurrying the soil.. During the time it takes for the NH_3/solvated electron solution diffuse into the soil (or for intercalated toxic compounds to extract into the NH_3), competitive reactions of solvated electrons with H_2O or NH_3 (catalyzed by Fe^{3+} or dissolved O_2) occur.

Table 1. Destruction of PCBs in Soils[17,20,21,23] Using Na/NH_3 or Ca/NH_3[a]

Treatment	Soil Type	Pre-treatment PCB Level (ppm)	Post-treatment PCB Level (ppm)	Destruction Efficiency (%)
Na/NH_3	Sand, silt, clay	770	<2.0	>98.4
Na/NH_3[b]	Sand, silt	1250	<2.0	>99.8
Na/NH_3	Sandy, clay	810	<1.0	>99.8
Ca/NH_3[c,d]	Clay	2140	2.0	>99.8
Ca/NH_3[d]	Sandy	6200	1.6	>99.9
Ca/NH_3[d]	Organic	660	0.16	>99.9

[a] Pre-weighed soil samples (75-100g) were slurried 10-20 min. at room temperature in liquid NH_3 in a 1 liter reactor. About 300cc of liquid NH_3 containing 1.27 to 3.3% wt. Na was used. Single treatment. No more Na or Ca was added.

[b] Reduction conducted at –33 °C.

[c] Montmorillinite clay-rich soil from Starkville, MS.

[d] Each soil was partially dried and then 20% wt.% water was added in addition to the water left after partial drying.

How do These Dechlorinations Work?

Aromatic dechlorinations in solvated electron solutions proceed (see Scheme 1) by rapid transfer of the solvated electron (e_s^-) to the chlorinated aromatic ring to give a radical anion, **11**, where the added electron goes into the LUMO (a π^* orbital) of the aryl ring. This is followed by loss of chloride. The resulting aromatic radical, **12**, acquires a second electron to generate anion **13**, which is then protonated to give reduction product **14**. This process continues successively until all the chlorines are replaced to give biphenyl, **15**. Continued, though slower, reduction to phenyl cyclohexene, **16**, can occur depending on the conditions. Moisture in the soil and acidic soil functional groups provide $^+NH_4$ irons which efficiently transfer protons to intermediate anions.

Scheme 1

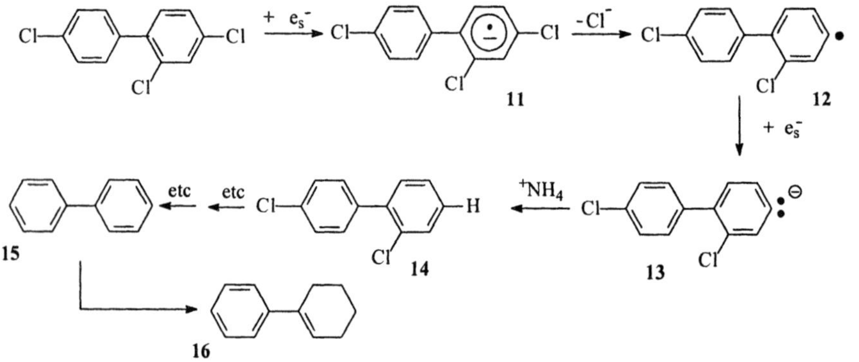

The transfer of a solvated electron to a chlorinated aliphatic hydrocarbon occurs with the simultaneous loss of chloride because there are no available π^* orbitals available.[32] Therefore, the electron must add to a carbon-chlorine antibonding sigma (σ) orbital, thereby breaking the C-Cl bond. This process is called dissociative electron transfer (see Scheme 2). The resulting radical can add a second solvated electron followed by loss of another chloride, as long as a chlorine is still present on that carbon. This is illustrated in Scheme 2 for the dechlorination of CCl_4 in Na/NH_3.

The reduction of CCl_4 was shown to proceed at diffusion controlled rates.[19] For example, when CCl_4 is treated with 2 or 3 equivalents of Na in NH_3 only CH_4 and CCl_4 were detected as products. No $CHCl_3$, CH_2Cl_2 or CH_3Cl were

Scheme 2

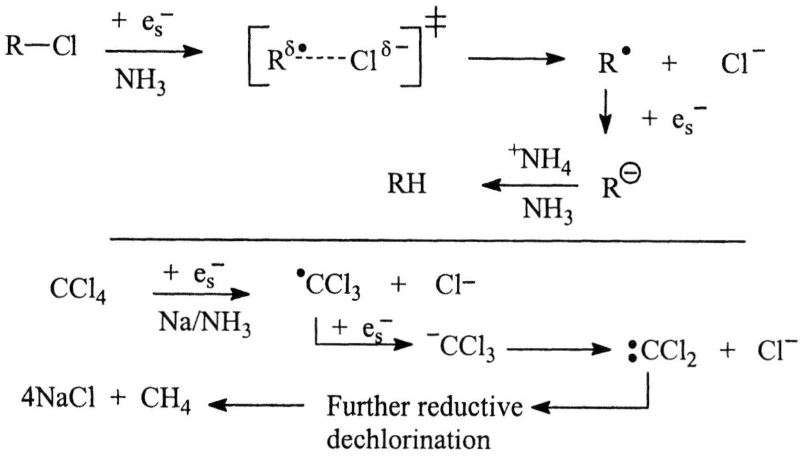

$$R-Cl \xrightarrow[NH_3]{+ e_s^-} [R^{\delta \bullet} \text{----} Cl^{\delta -}]^{\ddagger} \longrightarrow R^{\bullet} + Cl^-$$

$$\downarrow + e_s^-$$

$$RH \xleftarrow[NH_3]{^+NH_4} R^{\ominus}$$

$$CCl_4 \xrightarrow[Na/NH_3]{+ e_s^-} {}^{\bullet}CCl_3 + Cl^-$$

$$\xrightarrow{+ e_s^-} {}^-CCl_3 \longrightarrow :CCl_2 + Cl^-$$

$$4NaCl + CH_4 \longleftarrow \text{Further reductive dechlorination}$$

found. Thus, four reductions take place on the CCl_4 molecules in the vacinity of sodium particles (added to CCl_4/NH_3 solutions) before the partially dechlorinated intermediates can diffuse away. This was observed even when solutions were vigorously stirred.[19] Similar experiments were carried out with 3,4-dichlorotoluene.[19,22] Again, no monochlorotoluene was found. This was true whether or not water was present in the reductions (equation 1). By adding more sodium, complete reduction to toluene was achieved.

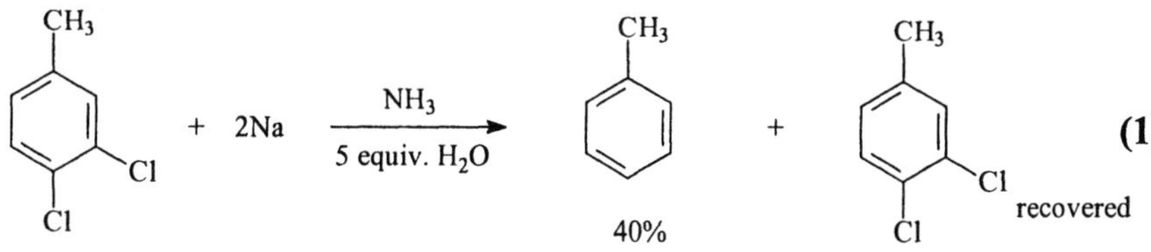

3,4-dichlorotoluene + 2Na $\xrightarrow[\text{5 equiv. } H_2O]{NH_3}$ toluene (40%) + 3,4-dichlorotoluene (recovered)   (1)

Why Do These Dechlorinations Work In The Presence of Excess Water?

Solvated electrons are destroyed at exceptionally fast rates in water. In pure water, the half-life of the solvated electron is only about $100\mu sec$![33,34] Furthermore, in a soil containing 20 wt.% H_2O and contaminated with 1000 ppm of PCB (using an average of 4 chlorines per PCB molecule) there are only 3×10^{-4} moles of PCB present for every mole of water. At 1.0 ppm of PCB there are only 3 X 10^{-7} moles of PCB per mole of water. In other words, water molecules outnumber PCB molecules by 3.3 X 10^3/1 and 3.3 X 10^6/1 at contamination levels of 1000 ppm and 1.0 ppm, respectively. At first glance, it seems impossible that solvated electrons could dechlorinate PCBs, or other chlorinated aromatics, without an enormous consumption of Na (or Ca or other alkali or alkaline earths). So how is it possible these remediations are feasible? The answer is based on two facts. First, in pure liquid NH_3 the half-life of a solvated electron is about 300h.[34] A portion of this stabilizing effect remains present even in 20% H_2O/80% NH_3, where the solvated electron's half-life is about 100 sec.[34,35] The second fact is that electron transfer to chlorinated organic compounds occurs very fast, hence dechlorinations are very fast. Therefore, the relative kinetics greatly favor dechlorination even though the mole ratios greatly favor water. The net effect is modest loss of the metal (Na, Ca ---) to side reactions with water.

The data summarized in Table 2[19,21,22] shows clearly that both aliphatic and aromatic model compounds can be reduced by Na/NH_3 without excessive consumption of Na when excess water is present. The minimum amount of Na required for complete dechlorination is express in two ways: (1) moles of Na/mole of chlorinated compound and (2) moles of Na/mole of chlorines present. This is shown for no water and for both 20 and 50 mole excesses of water. This data illustrates that only a modest increase in Na consumption occurred.

Soils and sludges with very high water contents can be partially dewatered before treatment. Also, if the contaminated matrix is first extracted by preslurrying in liquid NH_3, most of the chlorinated pollutants move into the NH_3 which is an excellent extractant. NH_3 nicely swells clays to allow intercalated PCBs, etc. to move into the solvent phase. Thus, upon addition of Na, the solvated electrons formed do not need to diffuse into clay layers or other soil aggregates prior to encountering the chlorinated pollutant. This reduces the time frame in which the reaction with water (equation 2) competes with reductive dechlorination. In over ten years of experience with soil and sludge remediations we have found remarkably little hydrogen was generated.

$$H_2O + e_s^- \longrightarrow H^\bullet + {}^-OH \longrightarrow \tfrac{1}{2} H_2 \qquad (2)$$

Extensive studies at Mississippi State University[17-23] and Commodor Solution Technologies[24-29] have indicated the efficiencies of Na and Ca are about equal when water is absent. However, the efficiency of Ca decreases relative to that of Na as the amount of water is increased. Neither Li nor K was as efficient as Na.[19,20] Laboratory reactions showed that Cl and Br are readily removed.[19,20] Aromatic fluorines are also readily removed unless the substrate is a phenol,

Table 2. Minimum amount of Na required to completely dechlorinate model compounds at 25 °C. Effect of added water.

Substrate	Na/Substrate Mole Ratio and (Na consumed per Cl Removed) Required for Complete Dechlorination. Effect of Water		
	H_2O/Substrate Mole Ratio		
	No H_2O	20/1	50/1
2-Chloro-*p*-xylene	1.5 (1.5)	2.4 (2.4)	2.5 (2.5)
1,2-Dichlorobenzene	2.8 (1.4)	4.5 (2.3)	5.0 (2.5)
1,2,3,4-Tetrachlorobenzene	5.0 (1.25)	7.0 (1.75)	8.6 (2.15)
2,4,6-Trichlorophenol	4.6 (1.53)	8.0 (2.66)	10.0 (3.33)
1,1,1-Trichloroethane	3.6 (1.2)	4.5 (1.5)	5.1 (1.7)
Carbon tetrachloride	4.6 (1.15)	5.5 (1.38)	6.4 (1.6)

where it was found that fluorine in the *para* position is removed very slowly.[20] Aliphatic fluoro compounds are not readily defluorinated relative to side reactions with water. Table 3 summarizes some representative results.

Example Remediation Studies

Many different soil contaminates have been treated including PCBs, CAHs, dioxins, furans, pesticides, chlorinated solvents, hexachlorobenzene etc. The New Bedford Harbor Sawyer Street site in Massachusetts is a superfund site due to PCB contamination of river sediments. Commodore Solution Technologies conducted a demonstration study where river sediment was washed with diisopropylamine (using the RCC B. E. S. T.™ process) which produced an oil concentrate containing PCB levels of 32,800 ppm. Dioxins/furans (TEFs) were also present at 47,000 ppt. This concentrate was treated with Na/NH_3 (Table 4). After treatment, the PCB level was 1.3 ppm and the dioxin/furans were also remediated. Thus, after treatment, the residue was well below regulatory requirements for disposal in non-hazardous waste landfills. Na/NH_3 treatment also removed lead, arsenic and selenium from the concentrate and these were recovered from the ammonia recycle unit.

Table 3. Activity of different metals and halides during solvated electron dehalogenations in liquid NH_3.

Substrate	H_2O/Substrate (mole ratio)	Metal/Substrate (Mole Ratio) Required For Complete Dehalogenation			
		Li	Na	K	Ca
2-Chloro-*p*-xylene	No H_2O	2.0	1.5	1.8	1.0
	50/1	6.0 (76%)[a]	2.5	5.7	2.5 (23%)[a]
		Halogen Substituent			
		F	Cl		Br
2-X-*p*-xylene	No H_2O	2.0	1.5		1.5
	50/1	4.0	2.5		2.5
4-X-phenol	No H_2O	2.0 (18%)[b]	2.0		2.2
	50/1		4.5		6.0

[a] Amount of substrate dechlorinated after rxn. was completed.
[b] Only 18% defluorination occurred in 20h.

Table 4. Na/NH_3 Treatment of PCB- and Dioxin-Contaminated Sludge from New Bedford Harbor, MA.

Contaminant	Pre-treatment (ppm)	Post-treatment (ppm)
PCB	32,800	1.3
Dioxin/Furan	47	0.012
Mercury	0.93	0.02
Lead	73	0.2
Selenium	2.5	0.2
Arsenic	2.8	0.1

Na/NH$_3$ was effective for destroying hexachlorobenzene in soils. Sandy soil containing 67.6 ppm of hexachlorobenzene from a site near Las Vegas, NV was treated with Na (4% wt.)/NH$_3$. The remediated soil contained less than 1 ppm of hexachlorobenzene. GC/MS analysis could not detect chlorinated products in the treated soil. Contaminated transformer oils and cutting fluids have been remediated using Na/NH$_3$ (See Table 5). Oils containing >20,000 ppm of PCBs were detoxified to levels below 0.5 ppm using NH$_3$ containing 2 to 4% wt. of Na. Just as impressive, Na/NH$_3$ successfully remediated dioxins present in waste oil from the McCormick and Baxter superfund site in California. These dioxin levels were reduced from 418,500 parts per trillion (ppt) to only 2.3 ppt and furans from 14,120 ppt to 1.3 ppt by a single treatment.

Table 5. Destruction of PCBs in Oils Using Na/NH$_3$

Oil	Temperature (°C)	Pre-treatment (ppm)	Post-treatment (ppm)
Motor Oil	16	23,339	<1.0
Transformer Oil	40	509,000	20[a], <2.0
Mineral Oil	40	5,000	<0.5
Hexane	40	100,000	0.5

[a] Na feed was deficient. Adding more Na completed the remediation.

Pesticides. The bulk destruction of 100 lb. quanties of malathion by Na/NH$_3$ has been carried out in Commodore Solution Technology's Marengo, OH facility. Nearly stoichiometric quantities of sodium were sufficient to give nondetectable levels of residual malathion. Most bulk samples of organic pesticides and hebicides containing halogens, phosphorous and sulfur are amenable to this treatment. Soils contaminated with the pesticides DDT, DDD, DDE and dieldrin from Hawaii and Virgina, were remediated to nondetectable levels on site at Port Hueneme, Ca. Naval station using a mobile demonstration unit. These results are summarized in Table 6.

The facile dechlorination of the "corner" chlorines of the pesticide Mirex, **2**, is of particular interest. These chlorines cannot be removed by nucleophilic substitution because nucleophiles cannot get inside the cage to approach carbon from the backside in S$_N$2 reactions. Elimination reactions are also impossible since double bonds from these corner carbons can't be formed due to the constrained geometry. However, solvated electron solutions readily and completely dechlorinate Mirex.[21]

Table 6. Remediation of Pesticides-Contaminated Soils Using Na/NH$_3$ in a Mobile Unit

Soil from:	DDT	DDD	DDE	Dieldrin
		(levels in ppm)		
Barbers Point, HI				
pre-treatment	200	180	69	-
post-treatment	< 0.02	< 0.02	< 0.02	-
Dahlgren, VA				
pre-treatment	9	1.6	-	15
post-treatment	< 0.02	< 0.02	-	< 0.02

CFCs. The phase out of Class I ozone-depleting compounds under terms of the Montreal Protocol drew attention to methods to dispose of large quantities of chlorofluorocarbon refrigerants and halons. Abel and Mouk[26,27,35] patented a process in which CFC feed-streams were treated with solvated electron/NH$_3$ solutions to achieve destruction efficiencies equal to or greater than the United Nations' target of 99.99%. Destruction efficiencies of 99.99% or greater were achieved at 18 °C using a 2:1 mole excess of Na (1.2 wt. % in NH$_3$) on 2.4 Kg quantities of bulk neat samples of such CFCs and Halons as CCl_3F, CCl_2F_2, CH_2F_2, CCl_2FCClF_2, CH_2FCF_3, $CBrClF_2$ and $CBrF_2CBrF_2$.

Reductive Remediation of Nonhalogenated Molecules. Na/NH$_3$ treatments can also destroy nonhalogenated hazardous compounds. Three classes pollutants will be mentioned here: polynuclear aromatic hydrocarbons (PNAs), nitro- and nitrate-type explosive wastes and chemical warfare agents. The treatment of neat samples of PNAs leads to destruction efficiencies of 99.99% for many of these compounds including such examples as acenaphthene, benzo[a]anthracene, benzo[b]fluoranthene, benzo[g,h,l]perylene, chrysene, fluoranthene, fluorine, naphthalene and phenanthrene. With the exception of naphthalene and anthracene, complex product mixtures are formed. Radical anion formation followed by protonation occurs sequentially leading to dihydro, tetrahydro and further reduced products (see Scheme 3). Depending on the reaction conditions, dimerization of intermediate radicals can occur to give dimers in various states of reduction.

Na/NH$_3$ treatments of PNA-contaminated soils have been carried out by J. He in our laboratory.[37] Both wet and dry soils were examined when contaminated with individual model PNAs and also with PNA mixtures. The reduction of PNAs is slower that that of halogenated organic compounds. Therefore, side reactions, such as reaction with water (equation 2) or both Fe^{+3} or O_2 catalyzed reactions of solvated electrons with NH_3[38,39] (equation 3), compete seriously with PNA reduction. This leads to greater consumptions of Na and sometimes multiple Na additions are necessary.

Scheme 3

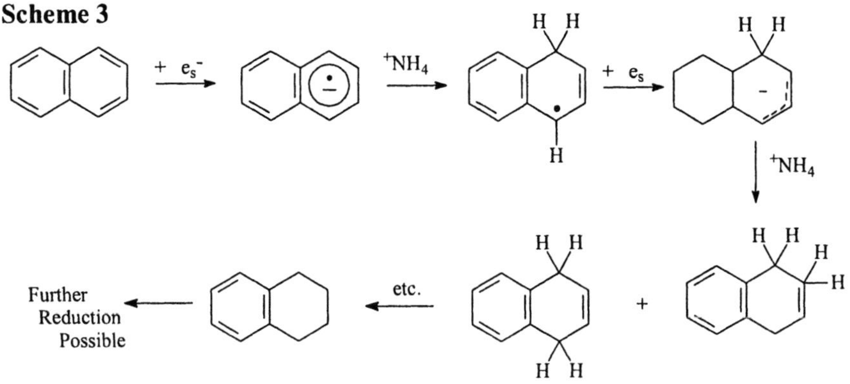

$$NH_3 + e_s^- \xrightarrow[O_2]{Fe^{+3} \text{ or}} ^-NH_2 + \tfrac{1}{2} H_2 \qquad (3)$$

Nitro- and nitrate containing organic compounds can also be reduced with solvated electron ammonia solutions. Aromatic nitro compounds may eventually be reduced to the corresponding amines but azoxy-, azo- and hydrazodimers were also obtained. Soils contaminated with explosives and propellant wastes are widespread. A number of explosives and propellants have been destroyed using Na/NH$_3$ including TNT, RDX, nitrocellulose, nitroglycerine, tetryl, PETN, Composition B and M-28. Neat explosives, samples from armaments and contaminated soils have all been treated. No explosive analytes were detected after analysis using EPA method 8330 (revision O, Sept. 1994). Example results from Na/NH$_3$ treatments of a RDX, HMX and 1,2-dinitrobenzene-contaminated soil from Los Alamos, NM are shown in Table 7.

Destruction of chemical warfare agents is another application for Na/NH$_3$ detoxification. Cost effective methods, other than combustion, are needed to dispose of aging stockpiles of chemical weapons. In extensive testing, the chemical agents GA, GB, GD, GF, Lewisite, VX, HD, HT, HN-1, HN-2, HL, picric acid, CG and CK have been destroyed with efficiencies greater that 99.9999% by Commodore. The reaction products have been characterized. The product mixtures were found to be CLASS 1 or CLASS 0 level materials when tested for acute toxicity. Scheme 4 illustrates the destruction of VX by rapid electron transfer to VX followed by 70% P-S and 30% C-S cleavage with evolution of ethane.

Solvated electron reductions can readily cleave P-OR, P-SR and P-halogen bonds as indicated both in Scheme 4 and by the destruction of GB. Nitrogen and sulfur mustard agents are also destroyed at diffusion controlled rates by Na/NH$_3$.[17,18]

Table 7. Na/NH$_3$ Remediation of Soil From Los Alamos, New Mexico Contaminated With Explosives. [a]

Treatment	HMX	RDX	1,2-DNB
Before Treating (mg/Kg)	1600	3580	9.6
After Na/NH$_3$ treatment [a]	0.03	0.03	0.03
Detection Limit	0.03	0.03	0.03
Destruction Efficiency (%)	>99.999	>99.999	>99.999

[a] Na (2.8 wt.%) in 1 liter of NH$_3$ was used per 50g of soil at 39 °C.

Scheme 4

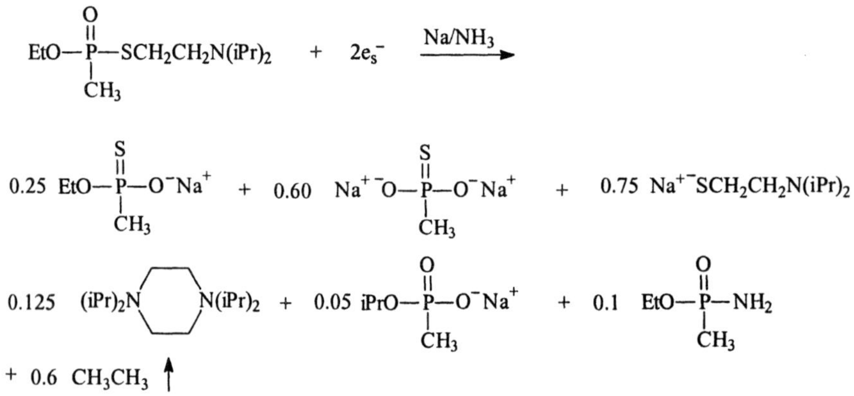

$$\text{EtO}-\overset{\overset{O}{\|}}{\underset{\underset{CH_3}{|}}{P}}-SCH_2CH_2N(iPr)_2 + 2e_s^- \xrightarrow{Na/NH_3}$$

$$0.25\ \text{EtO}-\overset{\overset{S}{\|}}{\underset{\underset{CH_3}{|}}{P}}-O^-Na^+ + 0.60\ Na^{+-}O-\overset{\overset{S}{\|}}{\underset{\underset{CH_3}{|}}{P}}-O^-Na^+ + 0.75\ Na^{+-}SCH_2CH_2N(iPr)_2$$

$$0.125\ (iPr)_2N\text{—}\underset{}{\bigcirc}\text{—}N(iPr)_2 + 0.05\ iPrO-\overset{\overset{O}{\|}}{\underset{\underset{CH_3}{|}}{P}}-O^-Na^+ + 0.1\ \text{EtO}-\overset{\overset{O}{\|}}{\underset{\underset{CH_3}{|}}{P}}-NH_2$$

$+ 0.6\ CH_3CH_3 \uparrow$

Defluorination of Aliphatic Fluoro Compounds

1-Fluorononane, a representative primary fluoroalkane, defluorinated slowly to nonane in Na/NH$_3$. After 15 min. at room temperature, only 39% fluorine cleavage occurred in a 12 mole excess of Na. Thus, fluoroalkanes would not compete effectively with the reaction of solvated electrons with water or with reduction of NH$_3$ to $^-$NH$_2$ in the presence of dissolved oxygen or Fe^{+3} which are present during remediations of environmental matrices. Aliphatic defluorinations are quite difficult due to the high bond strengths of sp^3-hybridized carbon-to-fluorine bonds. Therefore, our discovery[23,40] of the explosively fast and exothermic reaction between TiCl$_4$ and aliphatic fluorocarbons is of substantial interest. TiCl$_4$ exchanges chlorine into the alkane and rips fluorine away to form TiCl$_3$F. The reaction proceeds with more fluoroalkane to eventually produce TiClF$_3$ and 3 RCl. This rapid F/Cl exchange proceeds through a cationic mechanism but it does not isomerize the linear hydrocarbon chain (see equation 4) in the reaction of either 1-fluorooctane or 1-fluorononane with TiCl$_4$. This fluorine-chlorine exchange can be sequentially coupled with Na/NH$_3$ treatment which reductively dechlorinates the resulting mixture of chloroalkanes to the corresponding alkane. In the case of fluorocyclohexane, (equation 5) only chlorocyclohexane is formed upon reaction with TiCl$_4$.

$$\text{CH}_3(\text{CH}_2)_7\text{F} + \text{TiCl}_4 \xrightarrow[\text{r.t.}]{\text{hexane}} \text{TiCl}_3\text{F} + \text{CH}_3(\text{CH}_2)_7\text{Cl} + 5\% \quad + \left\{ \begin{array}{c} \text{dichloro isomers} \end{array} \right\} 77\% \quad (4)$$

$$\text{C}_6\text{H}_{11}\text{F} + \text{TiCl}_4 \xrightarrow[\text{r.t.}]{\text{hexane}} \text{C}_6\text{H}_{11}\text{Cl} \xrightarrow{\text{Na/NH}_3} \text{C}_6\text{H}_{12} \quad (5)$$

$>99\%$

A Further Examination of Na Consumption Versus the Concentration of the Chlorinated Contaminant in Soil Remediations With Na/NH$_3$.

The amounts of Na consumption required to achieve specific reductions in the levels of contaminant were studied using different soil types, water levels, and volumes of NH$_3$ on different chlorinated pollutants.[21-23,41] Some representative results for remediations of the model CAH, 1,1,1-trichloroethane, **5**, and also of tetrachloroethylene, **6**, are shown in Tables 8 and 9 respectively. Two types of soils (1) a clayey, mixed themic type Tapludults soil (B1) and (2) an organic soil with 3.4% organic matter (B2) were purposely contaminated to a known level. Then Na/NH$_3$ treatments were performed on both dry samples and samples containing specific known levels of water. Starting with a specific contamination level, soil samples were treated with different quantities of Na. The soils were then analyzed and the sodium consumptions were calculated. By incrementally adding specific amounts of Na and analyzing the amount of CH$_3$CCl$_3$ or Cl$_2$C=CCl$_2$ remaining after each treatment, it also was possible to

Table 8. Remediation of CH_3CCl_3 in B1 (clay) and B2 (organic) Soils With Na/NH_3[a,b,c]

Soil Type[c]	Water (wt%)	NH_3 Vol. per gram soil (ml)	CH_3CCl_3 level in soil (ppm)		Na Consumed per Cl Removed
			Start	Finish	
B1	dry	3	3000	9.6	7.4
		10	3000	7.5	7.4
B1	7%	3	3000	133	15
		10	3000	131	15
		3	3000	9.5	22
		10	3000	2.2	22
B1	15%	5	3000	381	22
		10	3000	361	22
B2	dry	5	3000	1.7	5
	7%	5	3000	7.2	22
	15%	5	3000	4.1	22

[a] Different amounts of Na were added and the Na required to remediate from the starting level to the final level was obtained.

[b] The same amount of Na was added in entries 1 and 2 but that amount was different from that added in entries 3 and 4, etc.

[c] Reference 21 (footnote c of Table 4) gives detailed characterizations of these soils.

Table 8. Remediation of Tetrachloroethylene-Contaminated B1 and B2 Soils Using Na/NH_3

Soil Type	Water Content (wt%)	$Cl_2C=CCl_2$ level in Soil (ppm)		Na Consumed per Cl Removed
		Start	Finish	
B1	dry	3000	0.9	9
	7%	3000	0.8	26
	15%	3000	1.0	38
B2	dry	3000	4.0	9
	7%	3000	4.6	21
	15%	3000	2.6	27

follow the sodium consumption per amount of chlorine removed as the remediation proceeded to lower concentrations of the pollutant in the soil.

Examining Table 8 and 9 reveals that more sodium is consumed per chlorine mineralized as the water level goes up. Within the range studied, the amount of NH_3 used per unit weight of soil did not influence the sodium consumption noticeably. Other experiments were performed on 1-chlorooctane in B1 soil.[21]

Those studies showed that the Na consumption per chlorine removed increased from 7 upon remediating from 5000 to 870 ppm to 300 while remediating from 100 to 7.4 ppm and 650 while remediating from 100 to 1.0 ppm. The overall consumption of Na per mole of chlorine was, however, only 30 when the initial contaimination level of 5000 was remediated down to 0.6 ppm. This reflects the increasing competition from side reactions as the chlorinated contaminant's concentration continuously decreases. This increase in the amount of Na required per Cl removed is not surprising because 6.9×10^5 molecules of water are present for every molecule of 1-chlorooctane at a contamination level of 0.6 ppm in a soil with 5% water. This emphasizes just how remarkably selective Na/NH_3 reductions are for halogenated organic molecules despite the very high reactivity of solvated electrons towards many functional groups. While the consumption of Na per chlorine removed seems high during remediations of PCBs, CAHs (or other chlorinated toxic molecules) from soils and sludges at concentrations below 50 ppm, one must keep in mind that very few moles of chlorine need to be remediated in the pollutant range from 50 to 0.1 ppm. Thus, high Na/Cl consumption ratios in this concentration range does not cause a large addition to the overall consumption of Na or to the cost of remediating these toxics to low levels.

Closing Perspective

PCBs and other chlorinated aromatic compounds are distributed in soils and sludges at over 400 sites in the USA. CAHs occur as serious contaminants at 358 major hazardous waste sites in the USA and they migrate vertically through soils to form DNAPLs on aquifer bottoms. Nitro compound wastes abound around ammunition plants and nitration operations. Every state is represented in this problem. One example is DOEs Hanford site which has a groundwater carbon tetrachloride plume extending over 70 sq. miles. Many contaminated sites exist in the Gulf Coast region (Texas through Florida) where the largest concentration of chemical manufacturing plants in North America is located together with many DOD sites. Thus, a national need exists for both *ex-situ* and *in-situ* methods to rapidly destroy these pollutants in soils and sludges at ambient temperature before they migrate into groundwater.

The EPA's latest Toxics Release Inventory report indicated that pesticides comprised 1%, PCBs 12%, mercury and mercury compounds 36% and polynucleararomatic hydrocarbons 44% of the bioaccumulative releases to the environment in 2000.[41] A total of 12.1×10^6 pounds of bioaccumulative releases were reported out of a total 7.10×10^9 pounds of all toxic releases in the United States. The total is down from 7.7×10^9 pounds in 1999. These releases have decreased 48%, or 1.6×10^9 pounds, for chemicals since 1988.[41]

Thus, progress is being made in reducing the rates of release. However, the continuing release of toxicants further adds to the load already present in the environment and not yet remediated. Clearly, improvements in remediation progress are sorely needed. With this national picture in mind and extrapolating to the problems existing worldwide, there should be a beneficial role for solvated electron reductions to play in various environmental restoration nitches.

Acknowledgements

Support of this work was provided by a Star grant from the US Environmental Protection Agency (Grant No: GAD#R826180) and by the Department of the Interior, US Geological Survey via the Mississippi Water Resources Research Institute (Grant Nos.: HQ96GR026769-12 and HQ01GR0088).

References

1. Wentz, C. A., *Hazardous Waste Management*, in: Clark, B. J.; J. Morris, M., Eds.; Chemical Engineer Series, McGwar-Hill, New York: **1989**.
2. Exner, J. H. *Detoxification of Hazardour Waste*, Ann Arbor Science Books: Ann Arbor, MI **1982**, p. 185.
3. Baillod, R. C.; Lampartes, R. A.; Laddy, D. G. *Wet Oxidation of Toxic Organic Substances*, in: Proceedings of the Purdue Industrial Waste Conference: West Lafayette, IN, **1978**.
4. Biros, F. J.; Walker, A. C.; Medbery, A. *Bull. Environ. Contam. Toxicol.* **1970**, 5, 317.
5. Sugimoto, H.; Shigenobu, M.; Sawyer, D. T. *Environ. Sci. Technol.* **1988**, 22, 1182.
6. Epling, G. A.; Florio, E. M.; Bourque, A. J. *Environ. Sci. Technol.* **1988**, 22, 952.
7. Tabaei, S. M. H.; Pittman, Jr., C. U. *Tetrahedron Lett.* **1993**, 34 (20), 3263-3266.
8. Tabaei, S. M. H.; Pittman, Jr., C. U.; Mead, K. T. *J. Org. Chem.* **1992**, 57, 6669.
9. Zhang, S.; Rusling, J. F. *Environ. Sci. Technol.* **1993**, 27, 1375.
10. Roth, J. A.; Dakoji,S. R.; Hughes, R. C.; Carmody, R. E. *Environ. Sci. Technol.* **1994**, 28, 80.
11. Chuang, F.-W.; Larson, R. A.; Wessman, M. S. *Environ. Sci. Technol.* **1995**, 29, 2460.

12. Yang,C.-M.; Pittman, Jr., C. U. *Hazardous Waste and Hazardous Materials* **1996**, 13/4, 445-464.
13. Erickson, M. D. S.; Swanson, E.; Flora, Jr., J. D.; Hinshaw, G. D. *Environ. Sci. Technol.* **1989**, 23, 462.
14. Pittman, Jr., C. U.; Yang, C. *J. Hazardous Materials* **2001**, B82, 299-311.
15. Oku , A.; Ysufuku, K. K.; Dataoka, H. *Chem. Ind.* **1978**, 84, 1.
16. Davies, W. A.; Prince, R. G. H. *Process Saf. Environ. Prot.* **1994**, 72, 113.
17. Pittman, Jr., C. U.; Tabaei, S. M. H. *Emerging Technologies in Hazardous Waste Management*, in: Tedder, D. W., Eds.: Atlanta, GA 27-29 September **1993**; Vol. 11, pp. 557-560.
18. Pittman, Jr., C. U.; Mohammed, M. K.; Extended Abstracts 1, in: *Proceedings of the EC Special Symposium on Emerging Technologies in Hazardous Waste Management*: Birmingham, AL, 9-11 September **1996**, pp. 720-723.
19. Sun, G.-R.; He, J.-B.; Pittman, Jr., C. U. *Chemospere* **2000**, 41, 907.
20. Pittman, Jr., C. U., Technical Report, Project No. 1434-HQ-96-02679-21, Department of the Interior, April **2000**.
21. Pittman, Jr.,C. U.; He, J.-B. *J. Hazardous Materials* **2002**, 92, 51-62.
22. Pittman, Jr., C. U.; He, J.-B., Technical Completion Report, Project No. HQ96GR02679-21, Department of the Interior, April **2001**.
23. Pittman, Jr., C. U.; He J.-B.; Sun, G.-R., Preprints of Extended Abstracts, Division of Environmental Chem., 220th Amer. Chem. Soc. National Meeting: Washington, DC., 20-24 September, **2000**; Vol. 40 (2), 784-787.
24. Weinberg, N.; Mazer, D. J.; Able, A. E. US Patent 4,853040, 1 August **1989**.
25. Weinberg, N.; Mazer, D. J.; Able, A. E. US Patent 5,110,364, 5 May **1992**.
26. Abel, A. E.; Mouk, R. W. US Patent 5,559,278, **1996**.
27. Abel, M. W.; Abel, A. E.; Heyduk; Mouk, R. W. US Patent 5,602,295, **1997**.
28. Abel, A. E. US Patent 5,495,062, **1996**.
29. Abel, A. E. US Patent 5,516,968, **1996**.
30. Smith, H. *Chemistry in Non-aqueous Solutions*; Interscience Publishers, **1963**, pp. 8-24 and 151-201.
31. Kennedy, M. V.; Stojanivic, B.; Shuman, Jr., F. L. *Environmental Quality* **1972**, 1, 63.
32. Holm, T. *J. Am. Chem. Soc.* **1999**, 121, 515.
33. Gould, R. F. *Advances in Chemistry*; American Chemical Society, **1965**, 50.

34. Schindewolf, U.; Metal-Ammonia Solutions, Lagowski, J. J.; Sienko, M. J. Eds.: Butterworths, London, **1970**, pp. 199-218.
35. Crooks, R. M.; Bard, A. J. *J. Chem. Phys.*, **1987**, 91 (5), 1274.
36. Mouk, R. W.; Abel, A. E. US Patent 5,414,200, **1995**.
37. Unpublished studies, He. J.-B.; Pittman, Jr. C. U.
38. Eastham, J. E.; Larkin, D. R. *J. Am. Chem. Soc.* **1959**, 81, 3652.
39. Rabideau, P. W.; Wetzeland, D. M.; Young, D. M. *J. Org. Chem.* **1984**, 49, 1544.
40. Pittman, Jr. C. U.; He, J.-B. *Proceedings of the Thirty-First Mississippi Water Resources Conference*, **2001**, p. 75-84, 10-11 April, Eagle Ridge Conference Center: Raymond, MS.
41. He, J.-B. "Dehalogenations of Aliphatic and Aromatic Hydrocarbons by Solvated Electrons: PhD Dissertation, Mississippi State University, December **2000**.
42. Hileman, B. "Dioxins Now On Toxics Inventory" *Chemical and Engineering News*, **2002**, June 3, 10. Start typing here

Chapter 15

Comparison of Atrazine and Alachlor Sorption, Mineralization, and Degradation Potential in Surface and Aquifer Sediments

Sharon A. Clay[1,*], David E. Clay[1], and Thomas B. Moorman[2]

[1]Plant Science Department, South Dakota State Northern University, Plains Biostress Laboratory, Brookings, SD 57007
[2]National Soil Tilth Laboratory, Agricultural Research Service, U.S. Department of Agriculture, Ames, IA 50011
*Corresponding author: Fax: 605-688-4452

Atrazine and alachlor have been used for controlling weeds in crops. These herbicides and their degradation products also have been detected in surface and ground waters. The objective of this study was to determine atrazine and alachlor sorption and degradation potentials at four soil depths and in aquifer matrix materials. Sorption potentials decreased with sample depth but were similar between herbicides at each depth. Mineralization potential differed between herbicides with twice as much alachlor mineralized at each depth compared with atrazine. After 112-d incubation at 10 C, atrazine degradation products were not detected in sample extract, whereas alachlor degradation products were present in both soil and aquifer matrix samples. Alachlor degradation products detected in A-horizon soil included the ethanesulfonic acid (ESA) and oxanillic acid (OAA) forms of alachlor. In aquifer extracts, two degradation products were formed, one cochromatographing with OAA-alachlor and an unidentified product. The ESA degradation product of alachlor has been detected frequently in aquifers at higher concentrations than alachlor. Its presence may be due to transport through the soil profile since this product was not detected in incubated samples.

© 2004 American Chemical Society

Introduction

Atrazine [6-chloro-N-ethyl-N'-(1-methylethyl)-1,3,5-triazine-2,4-diamine] has been and continues to be one of the most used soil and foliar applied herbicides for control of broadleaf and grass weeds in corn (*Zea mays*) in the United States. About 33 million kg of atrazine were applied in 1997 (1). Alachlor [2-chloro-N-(2,6-diethylphenyl)-N-(methoxymethyl)acetamide] (Figure 1) also has been widely used in mid-western United States corn and soybean (*Glycine max*) production as a soil treatment for control of problem grass weeds. Prior to 1999, about 23 million kg of alachlor was applied annually in the United States (1). In 1994, other chloracetamide compounds, including metolachlor [2-chloro-N-(2-ethyl-6-methylphenyl)-N-(2-methoxy-1-methylethyl)acetamide](29 million kg used in 1997 and ranked number 2 behind atrazine usage) and acetochlor [2-chloro-N-(ethoxymethyl)-N-(2-ethyl-6-methylphenyl)acetamide] (15 million kg used in 1997 and ranked number 3 in soil herbicide usage), began to replace alachlor in the marketplace (1) (Figure 1).

These four herbicides have been detected in aquifers throughout the United States (2). The amount of each herbicide detected in lakes, streams, resevoirs, and aquifers has been estimated to be not more than 1.5% of the total amount applied annually (3, 4). The percentage of ground water samples containing these herbicides range from 0.09 for acetochlor to 30% for atrazine (2, 5). The maximum concentrations for individual detections range from 0.02 ug/L for acetochlor to 5.4 ug/L for metolachlor (2).

Each of these herbicides degrades in the environment. Each degradate formed differs in it's degradation rate and soil retention properties (5). Atrazine is degraded microbially to deethylatrazine (DEA; 2-amino-4-chloro-6-isopropyl-s-triazine) and deisopropylatrazine (DIA; 2-amino-4-chloro-6-ethylamino-s-triazine) and by chemical hydrolysis to hydroxyatrazine (HA; 6-hydroxy-4-ethylamino-2-isopropylamino-s-triazine) (6). The chloracetanilide herbicides degradation in soil is the result of either microbially mediated glutathione conjugation (7, 8) or by the hydrolysis of amine group (9). The glutathione conjugation reaction with subsequent cleavage of the tripeptide glutathione molecule yields many polar compounds including ESA (ethanesulfonic acid) and OAA (oxanillic acid) metabolites of each respective chloractetanilide herbicide (10) (Figure 2).

In aquifers of the midwestern U.S., some of these herbicide metabolites have been detected more frequently and at higher concentrations than the parent molecule (3). Researchers in Wisconsin reported that alachlor, metolachlor, and acetochlor were detected in 20 and 40% of monitoring and private wells tested, respectively, while only 2% of the municipal samples contained these herbicides (11). At the same time, the ESA metabolites of each compound were detected in over 90% of monitoring and private wells and 50% of the municipal wells. In addition, OAA metabolites were detected in about 80% of the monitoring/private

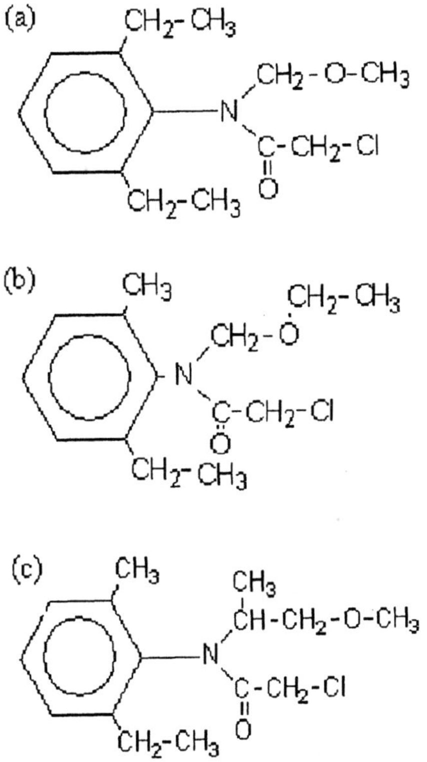

Figure 1. Chemical structures of (a) alachlor, (b) acetochlor, and (c) metolachlor.

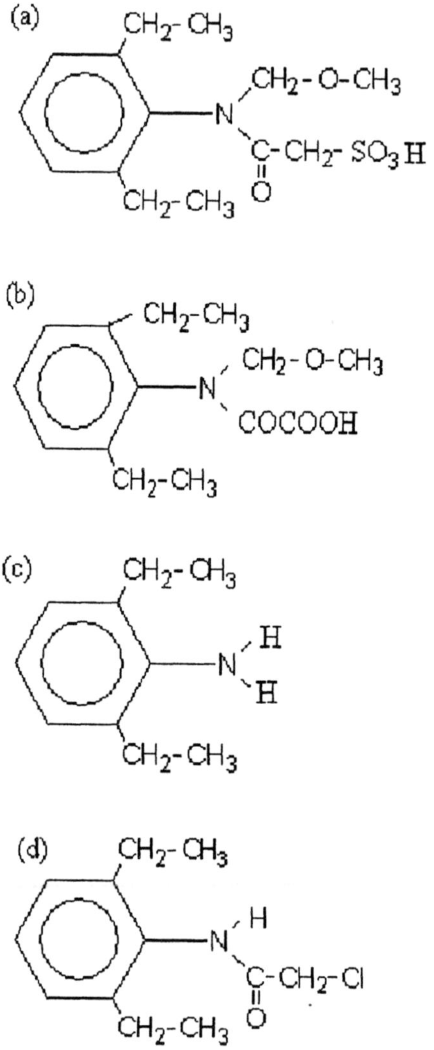

Figure 2. Chemical structures of selected alachlor metabolites, (a) ESA metabolite of alachlor, (b) OAA metabolite of alachlor, (c) 2,6-diethylaniline and (d) 2,6-diethylacetanilide.

wells and 40% of the municipal wells. In studies reported by Kolpin et al. (3), the ESA-alachlor metabolite was detected 10 times more frequently and at concentrations two times greater than alachlor. The ESA metabolites have higher water solubility than the original compounds (12) and it is thought that the relatively higher concentrations and greater detection frequencies of these metabolites may be the result of faster leaching through the soil profile. In contrast, the DEA metabolite of atrazine was found at the same frequency and concentration as atrazine while DIA was detected half as often and at half the maximum concentration of atrazine (3). The stability of DEA and atrazine in the environment may play a role in the frequency and maximum detection.

The toxicological properties of each herbicide are evaluated in stringent testing procedures. The toxicological properties of the herbicide degradates often are unknown due to the numerous and varied products that can be formed under the wide range of environmental conditions. There are several questions that are relevant. Are these major metabolites deactivated? Are these metabolites detoxified? Only a few of the metabolites have been assessed for toxicity and the methods vary widely due to the type of toxicity in question. For example, toxicity can be discussed in a number of different ways including acute toxicity, chronic toxicity, life-time exposure, oncogenicity, and genotoxicity. The toxicity rating can be evaluated on an array of species including plants, invertebrates, vertebrates, or human exposure. While these degradates may be less phytotoxic to both crop and weed species, their toxicity to other species may increase, decrease, or remain similar to the original herbicide.

Using the Microtox method of toxicity assesment that measures acute toxicity, DEA, DIA, and HA were reported to be less toxic to photobacterium than atrazine whereas the hydoxyalachlor metabolite had a toxicity rating similar to alachlor (13). The ESA-alachlor was found to pose little or no risk of producing adverse effects when assessing subchronic, genotoxic, and developmental toxicity to mice and rats (14). In further studies, ESA-alachlor was poorly adsorbed in the rat and those injesting ESA-alachlor showed no oncogenic responses, unlike the response to alachlor (15).

Aquifers of eastern South Dakota, western Minnesota, and northern Iowa are major sources of potable water for local communities. A high percentage of these aquifers are vulnerable to herbicide contamination due to: 1) their proximity to the soil surface (2 to 20 m below the surface); 2) intensive farming practices immediately above the aquifers; 3) soil types above the aquifers that are high in sand and/or low in organic matter; and 4) intense storm events where rainfall amounts can exceed 10 cm in a single storm. In studies conducted above the Big Sioux Aquifer in eastern South Dakota, atrazine was detected more often and at higher concentrations than alachlor although these two herbicides were applied at a similar application rates (16). However, alachlor was found much more frequently than atrazine in screening studies using monitoring wells across several eastern SD sites placed in several different aquifers (17).

Although alachlor is no longer used in the U.S., the three chemical compounds have very similar structural (Figure 1) and chemical properties. Alachlor degradataion data may be useful as a model for this chemical class. Caution must be used in interpolating these data however since the ESA metabolite of metolachlor is formed more slowly and at lower concentrations in soil (18). The objective of this study was to compare atrazine and alachlor sorption, mineralization, and degradation potential, processes that are major contributors to the environmental fate of pesticides, from surface soil to aquifer sediments in laboratory studies. In addition, degradation of alachlor was compared under aerobic and anaerobic conditions.

Materials and Methods

Surface and subsurface soil and aquifer matrix material (to an 8-m depth) were collected in an aseptic manner in presterilized 6.35 cm (width) PVC tubes using a hollow-stem auger from six sites over the Big Sioux Aquifer near Aurora, SD (19). The water table at the time was 7 m below the soil surface. The soil or matrix material was removed from the tubes in a sterilized laminar flow hood. The profile was divided into 6 zones that were designated as A (surface soil, from 0- to 0.3-m), B (soil from the B horizon, from 0.31 to 1.2-m), C1 (soil from the C horizon, from 1.4- to 3.5-m), C2 (soil from the C horizon, from 3.6- to 5.0-m), F (aquifer matrix material recovered just above the saturated zone, from 5.2- to 6.6-m) and S (aquifer matrix material from the saturated zone, from 6.8- to 8.8-m). All equipment used in this study was autoclaved or surface sterilized prior to use to prevent contamination. Bacterial counts were similar throughout the soil profile ranging from 8.47 to 7.98 log bacteria/g for the A and F zones, respectively (19). The number of fungal organisms, however, differed by orders of magnitude in the profile, from 6.81 to 1.28 log fungi/g for A and S zones, respectively.

Adsorption of atrazine and alachlor to soil and aquifer material was determined for each zone using batch equilibration techniques (19). To investigate alachlor degradation and mineralization, samples that consisted of 30 g of material were transferred into sterilized glass serum bottles. In order to achieve anaerobic conditions in selected treatments, the noncapped serum bottles containing sample were exposed to an anaerobic atmosphere (N_2 +CO_2 gas mixture) in an anaerobic glove box for at least 4 h or until equilibration occurred. The treatments included: 1) alachlor alone (aerobic and anaerobic conditions, all zones); 2) alachlor + carbon (aerobic and anaerobic conditions, all zones); 3) alachlor + nitrogen (aerobic and anaerobic conditions for F and S zones only); 4) alachlor +carbon + nitrogen (aerobic and anaerobic conditions,

A, F, and S zones). Alachlor was added at 16.6 pg/kg as a 1-ml aqueous aliquot labeled with 5 kBq of ^{14}C-uniformly-ring labeled herbicide.

Carbon was added to the herbicide aliquot as lyophilized algae biomass at a rate of 3.3 mg/kg. This mimicked the flush of dissolved organic carbon that has been detected during spring recharge of the aquifer (20). Nitrogen was added as the NO_3-N form at 0.8 mg N/kg. After treatment, sterilized water was added to all samples. Samples from the A to F zone were brought up to 30% water whereas samples from the S zone were brought up to a 1:1 soil/water ratio. Samples were capped and incubated at 10 C (the ambient temperature of the Big Sioux Aquifer). Each treatment for a given zone was replicated six times.

Aerobic samples were purged and aerated every 14 d up to 112 d, whereas anaerobic samples were purged only on 112 d. Compressed air was passed through a Drierite/Ascarite column to remove water and CO_2 and a silver filter to remove airborne microorganisms. The purged air from the sample was run through KOH traps to trap CO_2 and the amount of ^{14}C was quantified using liquid scintillation techniques.

After the final aeration, the remaining herbicide and degradates were extracted from the sample using 30 ml of 4:1 methanol/water solution. Methanol was removed by evaporation. The total amount of ^{14}C extracted was quantified in an aliquot by liquid scintillation counting. Another aliquot was spotted on thin layer chromatography plates and developed using butanol/acetic acid/water (6:2:3 v/v/v) to determine the amount of ^{14}C that remained as alachlor. Alachlor had a relative mobility (Rf value) of 0.75 and 88% of the ^{14}C-alachlor standard was detected in this band. Therefore, this band was scraped from each sample track to quantify the amount of alachlor remaining. Other Rf bands from selected treatments were scraped to provide information on other degradates. The Rf value of three degradates, 2'6'-diethlanaline, 2-chloro-2'6'-diethlacetanilide, and ethane sulfonic acid {[2-[(2,6-diethyl-phenyl)(methoxymethyl) amino]-2-oxoethanesulfonic acid}, were similar with an Rf value of 0.65. The oxoacetic acid metabolite (OAA) {[(2,6-diethylphenyl)(methoxymethyl) amino] oxoacetic acid} had an Rf value of 0.45 (Figure 2). Other bands of more polar, unidentified ^{14}C compounds were found in some treatments and had Rf values ranging from 0.05 to 0.35.

Atrazine mineralization was investigated in a very limited number of treatments. Aerobic samples from all zones were treated with atrazine at 16.6 pg/kg as a 1-ml aqueous aliquot labeled with 5 kBq of ^{14}C-uniformly-ring labeled herbicide. The samples were aerated as described above, and the amount of ^{14}C trapped was quantified.

Results and Discussion

Physical and chemical characteristics of the A horizon soil and aquifer matrix material have been reported in by Clay et al. (19). The soil in the A horizon had a silty clay loam texture (15% sand, 56% silt, and 27% clay). The C horizon was glacial outwash, classified as a sandy loam soil, with 90% of the silt particles made up of nonreactive quartz. Over 60% of the material collected from the F and S zones was gravel and rock materials >4 mm, the remaining smaller particles were similar in makeup to the silt particles in the C horizon. Organic carbon ranged from 2688 to 44 mg/kg in the A horizon soil and aquifer matrix material, respectively (Table 1). Atrazine sorption (Kd value) ranged from 5.4 L/kg in the A horizon to about 0.4 in the F and S zones (Table 1). Alachlor had Kd values that ranged from 5.8 (A horizon) to 0.4 (F and S zone material. Sorption in the F and S zones is most likely overestimated for both compounds since sorption studies were conducted on particles <2mm in size.

Atrazine mineralization, total breakdown of the triazine ring structure and determined by accounting for $^{14}CO_2$ collected in KOH traps, ranged from 1.1% in surface soil to 0.15% material from the C1 zone. Mineralization in materials from the F and S zones was about 0.6%. Extraction recovery of the ^{14}C from the soil and matrix samples averaged about 80% of the amount applied. TLC analysis of extracts indicated that no significant amounts of atrazine degradates were present.

Alachlor mineralization was estimated to be 2.5% in surface soil, 0.16% in material from the F zone, and 0.3% in the aquifer matrix material from the S zone. Extraction recovery of the ^{14}C after incubation averaged about 90% of the amount applied.

Alachlor was degraded in soil and aquifer matrix material. The amount of alachlor that remained under aerobic conditions after 112 d ranged from 30% in the B horizon soil to about 80% in the C2 soil (Figure 3). Under anaerobic conditions, alachlor degradation was much less, with 70% or more of the applied alachlor remaining after 112 d.

The addition of the algae biomass under aerobic condition increased alachlor degradation in the A horizon soil and the F and S sediments. At the end of 112 d in treatments with C addition, less than 2% of alachlor remained in the A horizon soil whereas about 30% alachlor remained in the F and 10% remained in the S sediments. Under anaerobic conditions, the addition of N increased degradation in the F and S sediments with about 70% remaining as alachlor (vs about 80% without N addition).

Several metabolites of alachlor were present under aerobic and anaerobic conditions (Table 2). The alachlor standard had an Rf value of 0.75. In A horizon soil under aerobic conditions, about 62% of the radioactivity cochromatographed with alachlor. Compounds with Rf values on 0.65 (which cochromatographed with 2', 6'diethylanaline, 2-chloro-2', 6'-diethalacetanilide, and ESA metabolite), 0.55, and 0.5 were observed also. With the addition of N, additional radioactive compounds were detected with Rf values of 0.45 (that cochromatographed with OAA metabolite) and 0.35. Under anaerobic conditions, radioactive compounds were found at 0.7, 0.6, 0.5, 0.35, and 0.15.

Profile designation	Depth from soil surface (m)	Organic carbon (mg/kg)	K_d (L/kg) Atrazine	K_d (L/kg) Alachlor
A horizon	0-0.3	2688	5.4	5.8
B horizon	0.3-1.2	100	0.1	1.9
C horizon	1.4-3.5	20	0.2	-
Fringe sediment[a]	5.2-6.6	44	0.4	0.4
Saturated sediment[a]	6.8-8.0	45	0.4	0.4

[a]The values for the fringe and saturated sediments are for materials <2 mm in size. About 60% of the material from these samples was >4 mm in size.

Table 1. Organic carbon content and partition coefficients (K_d) of atrazine and alachlor by depth for the Brandt silty clay loam soil and aquifer sediments.

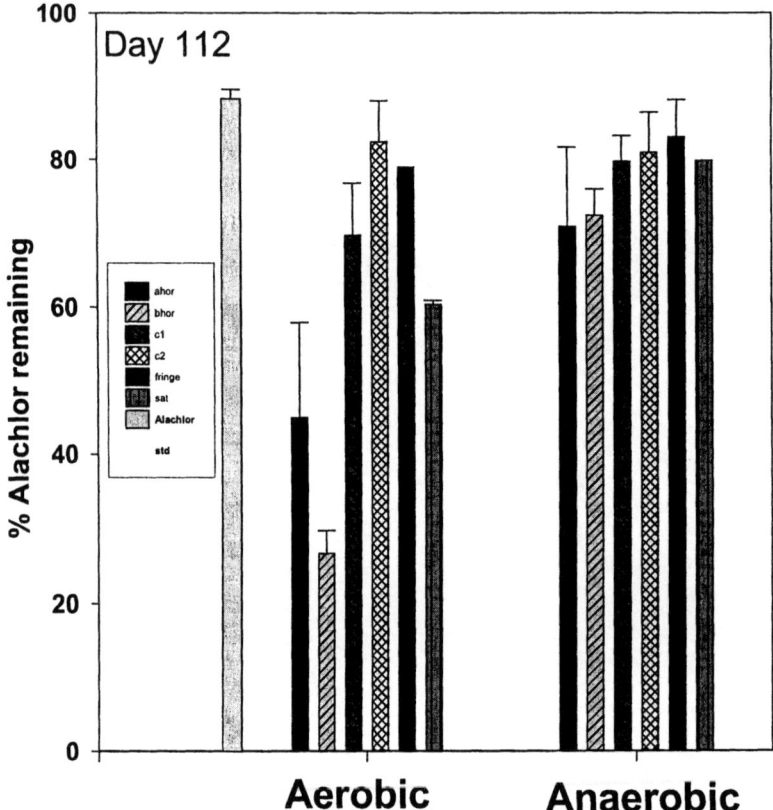

Figure 3. Alachlor remaining after 112 d under aerobic and anaerobic conditions incubated at 10 C in soil and aquifer matrix material taken from the Big Sioux Aquifer region of South Dakota near Aurora, SD.

Compound Rf value	A horizon soil			Fringe sediment				Saturated sediment	
	Aerobic		Anaerobic	Aerobic			Anaerobic	Aerobic	Anaerobic
	untreated	+N	untreated	untreated	+N	+N+C	untreated	untreated	untreated
0.75 (alachlor)	+	+		+	'+'	'+'	'+'	+	+
0.70			+	+	s	s			
0.65 (a, d)	+	+							
0.60			+				+	+	
0.55	+	+	+	+	s	s			
0.50	+	+	+						
0.45 (b)		+						+	
0.35		+	+						
0.15			+						
0.05							+		

Table 2. Radioactive regions (+) of thin layer chromotographic plates. The letter 's' indicates that a lesser amount of this compound was detected than when the sediment was untreated. Extracts were obtained from soil and aquifer sediment 112 d after alachlor treatment (incubation temperature of 10 C). Letters behind the Rf value refer to alachlor degradation products shown in Figure 2. Another alachlor compound that cocomatagraphed to 0.65 was 2-chloro-2',6'-diethalacetanilide.

In the F sediment under aerobic conditions, alachlor and compounds with Rf values of 0.7 and 0.55 were observed. Adding N or N+C, reduced the amount of radioactivity in these bands with more found as alachlor. Under anaerobic conditions, additional compounds were detected at 0.6 and 0.05. In sediment from the S zone, under saturated, aerobic conditions, compounds that cochromatographed with alachlor plus compounds at 0.6 and 0.45 were observed. Under anaerobic conditions, only the band that cochromatographed with alachlor was observed. The detection of ESA-alachlor in sediments under anaerobic conditions via the glutathione-s-transferase enzyme would not be expected since oxygen for the oxidation of the glutathione-derived sulfur would be absent.

Locke et al. (21) reported that after 54 d of incubation in surface soil at 25 C, the ESA and OAA were the major polar extractable metabolites found. In our study, about 13% of the alachlor applied to saturated sediment was found in polar metabolites. Adding C + N increased the amount of polar metabolites to about 30%.

Potter and Carpenter (22) reported that once an acetanilide herbicide is in the aquifer, it persists for a relatively long time. The sorption and degradation of alachlor in surface and aquifer sediments are similar to results reported for much deeper aquifers (23). However, we estimated the half-life of alachlor in this shallow aquifer to be about 280 d, whereas the half-life from the deep aquifer samples reported by Pothuluri et al. (23) was estimated to be about 320 d. Radosevich et al. (24) reported degradation in only 4 out of 83 aquifer samples and Novick et al. (8) reported no mineralization of alachlor in aquifer sediment, although polar metabolites were formed. The addition of C increased alachlor degradation under aerobic conditions, indicating that degradation was C limited and is consistent with reports about the cometobolic nature of alachlor degradation.

Alachlor was degraded more slowly and fewer (if any) metabolites were produced under anaerobic conditions than under aerobic conditions. This finding may reflect reduced microbial activity under anaerobic conditions or the reduced enzymatic capacity of the microorganisms that are active under anaerobic conditions. The addition of nitrate to the sediments increased alachlor mineralization, suggesting that electron acceptors could be limiting. Stamper et al. (25) reported alachlor degradation under sulfate reducing conditions in surface soils and pond sediments.

Increased herbicide persistence under anaerobic conditions probably increases the risk of increased transport in groundwater. The detections of nitrate in this aquifer system (20) suggests that aerobic or denitrifying conditions are probably more prevalent than more strongly reduced conditions in this groundwater system.

References

1. Aspelin A.L.; Grube A.H. *Pesticides Industry Sales and Usage: 1996 and 1997 Market Estimates.* EPA/OPP. http://www.epa.gov/oppbead1/pestsales/97pestsales. **2000; (accessed April 2002).**
2. Kolpin, D.W.; Barbash, J.E.; Gilliom, R.J. 2000. Pesticides in ground water of the United States, 992-1996. *Ground Water* 2000, *38*, 858-863.
3. Koplin, D.W.; Thurman, E.M.; Goolsby, D.A. Occurrence of selected pesticides and their metabolites in near-surface aquifers of the midwestern United States. *Environ. Sci. Technol.*, 1996, *30*, 335-340.
4. Larson S.J.; Gilliom R.J.; Capel P.D. *Pesticides in Streams of the United States- Initial Results from the National Water-Quality Assessment Program.* USGS/NAWQA, WRIR98-4222, RLINK http://www.water.wr.usgs.gov/pnsp/rep/wrir984222. **1999**, (accessed April 2002).
5. Koplin D.W.; Nations B.K.; Goolsby D.A.; Thurman E.M. Acetochlor in the hydrologic system in the midwestern United States, 1994. *Environ. Sci. Technol.*, **1996,** *30*, 1459-1464.
6. Barrett, M.R. The enviromental impact of pesticide degradates in groundwater. In: M.T. Meyer and E.M. Thurman (eds) Herbicide Metabilites in Surface Water and Groundwater. ACS Symp Ser. 630. American Chemical Society, Washington, DC, **1996;** p 200-225.
7. Lamoureux, G.L.; Rusness, D.G. Propachlor metabolism in soybean plants, excised soybean tissues, and soil. *Pestic. Biochem. Physiol.* **1989**, *34*, 187-204.
8. Feng, P.C.C. Soil transformation of acetochlor via glutathione conjugation. *Pestic. Biochem. Physiol.* **1991**, *0*, 6-142.
9. Novick, N.H.; Mukherjee, R.; Alexander, M. Metabolism of alachlor and propachlor in suspensions of pretreated soils and in samples from groundwater aquifers. *J. Agric. Food Chem.* **1986**, *34*, 721-725.
10. Sharp, D. Alachlor. In: P.C. Kearney, D.D. Kaufman (eds). *Herbicides: Chemistry, degradation, and mode of action. Vol 3.* Marcel Dekker, New York, **1988**; 301-333
11. Rheineck, B. and J. Postle. Chloroacetanilide Herbicide Metabolites in Wisconsin Groundwater Final Report. 9 pgs. Wisconsin Department of Agriculture, Trade and Consumer Protection. **2001.** www.soils.wisc.edu/extension/FAPM/proceedings01 (Accessed June, 2002).
12. Thurman, E.M.; Meyer, M.T. Herbicide metabolites in surface and groundwater: Introduction and review. *In*: Meyer, M.T.; Thurman, E.M. (ed.) Herbicide metabolites in surface and groundwater. ACS Symp. Ser. 630 Am. Chem. Soc., Washington, DC, **1996,** p. 1-15.

13. Kross, B.C.; Vergara, A.; Raue, L.E. Toxicity assessment of atrazine, alachlor, and carbofuran and their respecive environmental metabolites using Microtox. *J. Toxicol. Environ Heath.* **1992.** *37*, 149-159.
14. Heydens, W.F.; Siglin, J.C.; Holson, J.F.; Stegeman, S.D. Subchronic, developmental, and genetic toxicology studies with ethane sufonate metabolite of alachlor. *Fundam. Appl. Toxicol.* **1996,** *33*, 173-181.
15. Heydens, W.F., A.G.E. Wilson, L.J. Kraus, W.E. Hopkins III, and K.J. Hotz. Ethane sulfonate metabolite of alachlor: Assessment of oncogenic potential based on metabolic and mechanistic considerations. *Toxicological Sci.* **2000,** *55*, 36-43.
16. Clay, S.A.; Clay, D.E.; Koskinen,W.C.; Berg, Jr., R.K. Application method: impacts on atrazine and alachlor movement, weed control, and corn yield in three tillage systems. *Soil Tillage Res.* **1998,** *48,* 215-224.
17. Davis, J.R. Analysis and evaluation of field monitoring. In *Oakwood Lakes-Poinsett: Rural Clean Water Program Comprehensive Monitoring and Evaluation,* Technical Report Project 20. Institute of Water Resources, South Dakota State University, Brookings, SD. **1989.**
18. Aga, D.S.; Thurman, E.M. Formation and transport of the sulfonic acid metabolites of alachlor and metolachlor in soil. *Environ. Sci. Technol.* **2001,** *35*, 2445-2460.
19. Clay, S.A.; Moorman, T.B.; Clay, D.E.; Scholes, K.A. 1997. Sorption and degradation of alachlor in soil and aquifer material. *J. Environ. Qual.* **1997,** *26,* 1348-1353.
20. Clay, D.E.; Clay, S.A.; Moorman, T.B.; Brix-Davis, K.; Scholes, K.A.; Bender, A.R. Temporal variability of organic C and nitrate in a shallow aquifer. *Water Res.* **1996,** *30,* 559-568.
21. Locke, M.A.; Gaston, L.A.; Zablotowicz, R.M. Alachlor biotransformation and sorption in soil and two soybean tillage systems. *J. Agric. Food Chem.* **1996,** *44,* 1128-1134.
22. Potter, T.L.; Carpenter, T.L. Occurrence of alachlor environmental degradation products in groundwater. *J. Environ. Qual.* **1995,** *29,* 1557-1563.
23. Pothuluri, J.V.; Moorman, T.B.; Obenhuber, D.E.; Wauchope, R.D. Aerobic and anaerobic degradation of alachlor in samples from a surface-to-groundwater profile. *J.Environ. Qual.* **1990,** *19*, 525-530.
24. Radosevich, M.; Crawford, J.J.; Traina, S.J.; Oh, K.-H.; Tuovinen, O.H. Biodegradation of atrazine and alachlor in subsurface sediments. *In* D.M. Linn, D.M. et al. (ed.) Sorption and degradation of pesticides and organic chemicals in soil. SSSA Spec. Publ. 32. SSSA, Madison, WI. **1993.** p. 33-43.
25. Stamper, D.M.; Traina, S.J.; Tuovinen, O.H. Anerobic transformation of alachlor, propachlor and metolachlor with sulfide. *J. Environ. Qual.* **1997,** *26*, 488-494.

Chapter 16

Pesticide Runoff and Mitigation at a Commercial Nursery Site

J. N. Kabashima[1], S. J. Lee[2], D. L. Haver[1], K. S. Goh[3], L. S. Wu[2], and J. Gan[2]

[1]University of California Cooperative Extension, Orange County, 7601 Irvine Boulevard, Irvine, CA 92618
[2]Department of Environmental Sciences, University of California, Riverside, CA 92521
[3]California Department of Pesticide Regulation, 1001 I Street, Sacramento, CA 95812

Nursery production plays a critical role in the modern agricultural economy. However, currently water quality issues related to pesticide and nutrient runoffs are exerting great pressures on many nurseries. The continued prosperity of the nursery industry will rely on adequate compliance with such environmental regulations as nonpoint source permits and total maximum daily loads (TMDLs). In close collaboration with nursery growers, we have carried out studies to understand the fate and distribution of pesticides in nursery runoff, and to develop best management practices (BMPs) to reduce pesticide load in the runoff. The experimental site was a 100-acre commercial nursery located in southern California. Two synthetic pyrethroids, bifenthrin and permethrin, were monitored in the runoff. The BMPs included optimized irrigation schemes, use of sediment traps/ponds, addition of polyacrylamide (PAM) into the effluent, and establishment of a vegetative strip. Monitoring data showed that the BMPs were highly effective in reducing the runoff of the synthetic pyrethroids. These BMPs are inexpensive and of low maintenance, and therefore are feasible for implementation at other runoff sites.

Introduction

Nursery production is an important industry in California. To maintain plant vigor and meet quarantine requirements, pesticides and fertilizers are used heavily in nursery production. These uses, when coupled with frequent irrigation, often lead to runoff and offsite movement of pesticides and nutrients. Because many nurseries are situated in urban or suburban environments, uncontrolled nursery runoffs generally enter nearby urban streams and eventually enter large creeks or ocean estuaries. For example, California Department of Pesticide Regulation (CDPR) monitoring study showed a number of insecticides, including both organophosphate insecticides and synthetic pyrethroids, in nursery runoff in Orange County (1). Due to the high toxicity of organophosphate and synthetic pyrethroid insecticides to fish and aquatic invertebrates, regulatory agencies are imposing stringent limits for pesticides in surface water. Nursery runoff is also expected to contribute to the overall pesticide load in urban streams. For instance, diazinon and chlorpyrifos, both organophosphate insecticides, were detected in the Newport Bay Watershed in Orange County at concentrations that were well above the numeric target values (2). Consequently, EPA recently adopted several Total Maximum Daily Loads (TMDLs) for diazinon and chlorpyrifos in the Newport Bay Watershed. TMDL implementation requires adoption of management practices for load reduction. Nursery runoff problems in California are not confined to Orange County. Due to runoff concerns, nursery and greenhouse operations in many other regions in California are also under great public pressure. Nor is runoff confined to nurseries alone. Pesticide runoff has also been observed in orchards (3, 4), golf courses (5), urban residential landscapes (2), and other agricultural systems (6, 7). Therefore, controlling runoff to meet water quality standards is a challenge faced by many sectors of urban and agricultural developments.

A mid-size commercial nursery in Orange County, California, was selected as the field site in 1999 by researchers from both CDPR and University of California for monitoring pesticide runoff and developing risk-mitigation practices. The nursery occupies about 100 acres, and has been in operation since 1958. The nursery production includes greenhouses, lathe houses, and outdoor areas that are padded with various impermeable or semi-permeable materials. In greenhouses and lathe houses, plant containers are typically irrigated through drip irrigation. In outdoor areas, plant containers are irrigated with overhead sprinklers. It has been observed that runoff is generated from all these areas, under both drip and sprinkler irrigation practices.

Since 1998, as a component of the quarantine practices against red and imported fire ant (RIFA) infestation, all plant containers at this and other nurseries in this region have been required to be treated with bifenthrin (2-methylbiphenyl-3-yl-methyl(Z)-(1*RS*)-*cis*-3-(2-chloro-3,3,3-trifluoropropenyl)-2,2-dimethylcyclopropane-carboxylate). The treatment is typically carried out by incorporating a granular bifenthrin formulation, Talstar®, into potting mixes before planting or seeding. As part of the RIFA quarantine program, California Department of Food and Agriculture (CDFA) and CDPR started a monitoring study to determine the possible presence of RIFA-mandated insecticides at several nursery discharge points and in surface streams that receive the runoff. Many insecticides, including bifenthrin and organophosphate insecticides, display relatively high acute toxicity to fish and indicator aquatic invertebrates (Table 1). For instance, LC_{50} of bifenthrin for *Ceriodaphnia dubia* is 0.078 ppb, or 78 ppt.

Nursery production is a unique system in terms of pesticide fate and safety. Currently little is known about the environmental fate of pesticides in plant containers or in adjacent environments. Nursery sites are heterogeneous in terms of surface conditions and water management, and production activities vary greatly in different seasons and even on different days. Runoff paths at nursery sites and small streams that initially receive the runoff represent environmental settings that are very different from systems such as agricultural fields. In particular, due to constant production activity, the low density of potting mixes, and the relatively impervious surfaces at most nursery sites, nursery runoff is characterized by high levels of suspended solids composed of silt, clay, and organic matter particles. Therefore, pesticides can conceptually move in nursery runoff in two forms: adsorbed to suspended solids and dissolved in water. The association with suspended solids is especially important for understanding the runoff behavior of the strongly adsorbing pesticides, such as bifenthrin. This characteristic has been found also important for designing risk-management practices aiming to remove bifenthrin and similar pesticides from the runoff.

Two synthetic pyrethroid insecticides, bifenthrin and permethrin (3-phenoxybenzyl (1*RS*)-*cis-trans*-3-(2,2-dichlorovinyl)-2,2-dimethylcyclopropane-carboxylate), are the major runoff constituents considered in our monitoring and evaluation studies. Bifenthrin has a solubility of <1 ppb, and a K_{oc} of 240,000 ml g^{-1}. Permethrin has a solubility of 6 ppb in water, and a K_{oc} of 100,000 ml g^{-1} (*8*). The strong adsorption on soil suggests that a large fraction of the pesticide is associated with the suspended solids in nursery runoff. This determines that the downstream distribution of these pesticides is closely related to distribution of sediments, and that practices aiming to remove suspended solids from runoff will also be effective in reducing the level of these pesticides.

Experimental

Site Characterization and Survey for Pesticide Sources

The nursery is located in the upper part of the San Diego Creek/Newport Bay Watershed in Orange County, CA, and the runoff drains through an artificial channel and eventually into the San Diego Creek. At the nursery site, many small

Table 1. LC50 values (ppb) of insecticides used for control of red and important fire ants (RIFA)

Insecticide	R. trout	D. magna	C. dubia
Bifenthrin	0.15	0.16	0.078
Chlorpyrifos	3	1.7	0.13
Diazinon	2600	0.96	0.51
Fenoxycarb	1600	400	na
Hydramethylnon	160	1140	na
Malathion	170	1.8	1.14
Pyriproxyfen	>325	400	na

runoff streams that are generated in various production areas converge into two main runoff paths, which further flow into a settling pond. The runoff then flows through a 260-m long cement-lined channel before it leaves the property. The runoff paths are depicted schematically in *Figure 1*.

To understand the source of bifenthrin and permethrin in the runoff, we collected surface soil samples at selected locations at the nursery and analyzed bifenthrin and permethrin levels in the samples. The sampling locations are shown in *Figure 2*. Loose soil particles were gathered from a 50 × 50 cm area using a brush and collected in plastic bags. The sample sites were located next to various runoff paths. The soil at most sites contained wood chips that originated from potting soil mix; the percentage of each soil sample comprised of potting mix was estimated based on a visual assessment of wood chip content. The widespread distribution of potting soil was a result of potting mix spills during transportation of plant containers and movement in surface runoff.

Aliquots (2~5 g) of the soil samples were extracted by shaking with 30 ml hexane-acetone (1:1, v/v) in 50-ml Teflon-coated centrifuge tubes for 1 h. The mixtures were then centrifuged at 10,000 rpm in a centrifuge and a subsample of the extract was transferred to an autosampler vial for analysis using the method as described below.

Best Management Practices (BMPs)

A number of practices have been implemented over the last few years to reduce runoff output and pesticide load in the runoff. First of all, irrigation systems were computerized to optimize irrigation interval and time, and as a result, the overall runoff output has gradually decreased, as evident from the

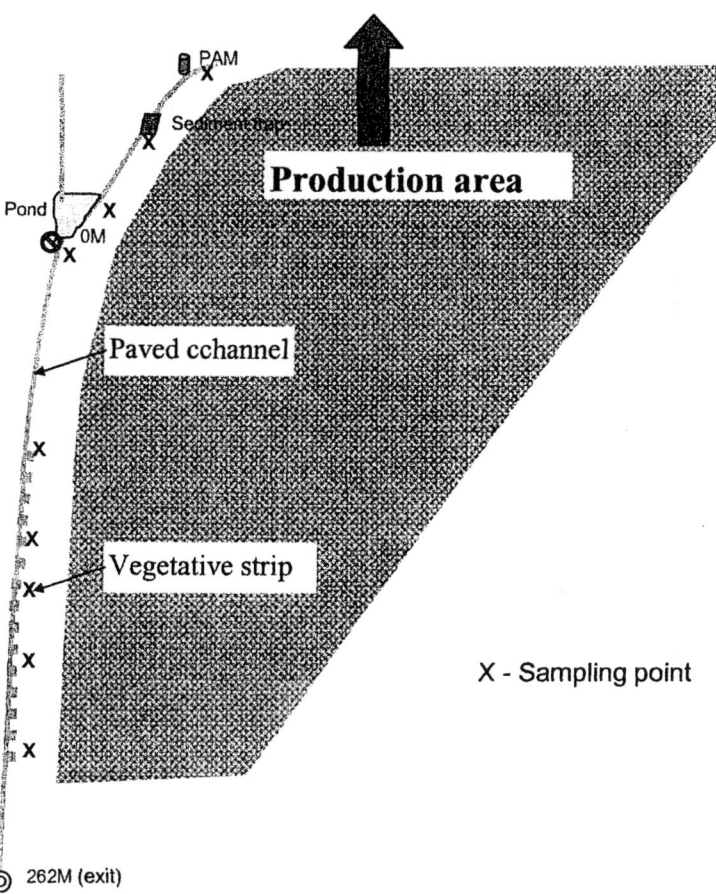

Figure 1. Arrangement of best management practices (BMPs) and location of sampling points

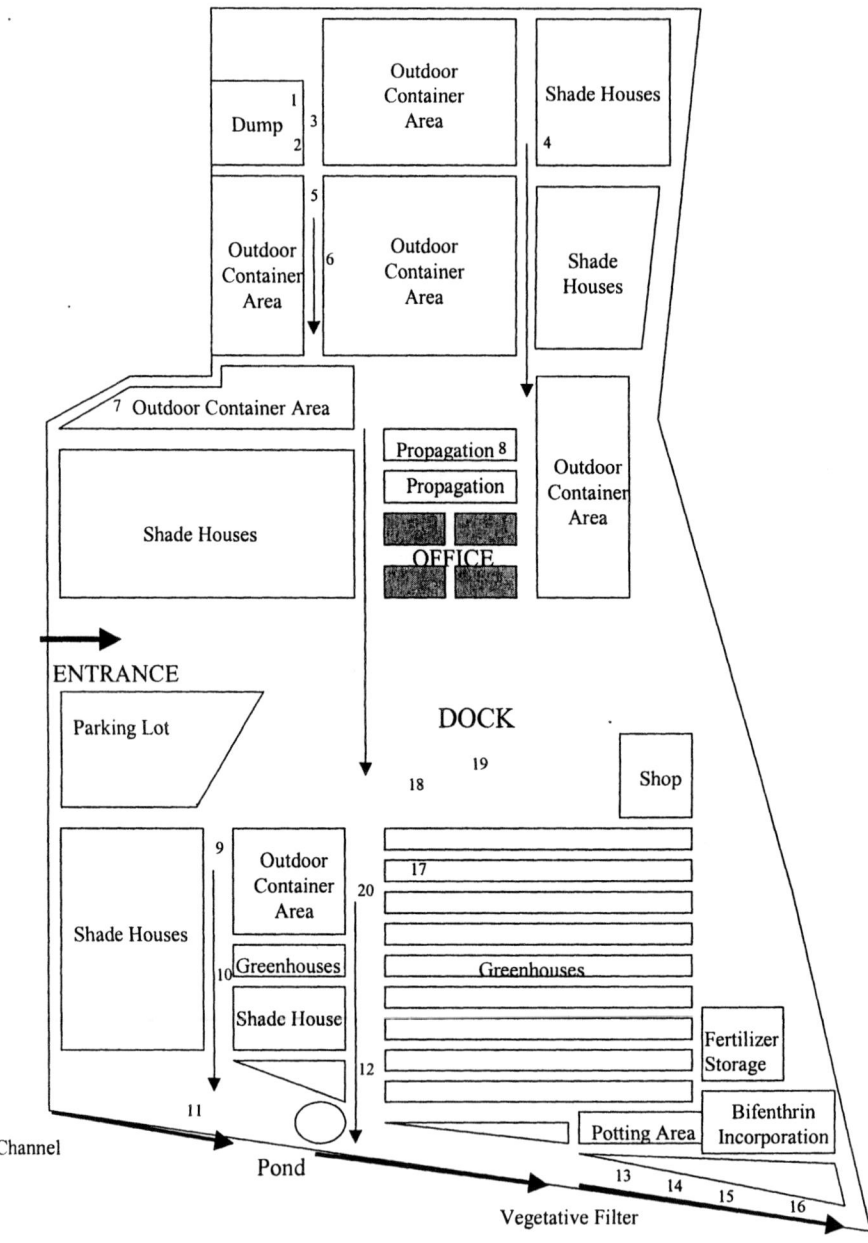

Figure 2. Locations for surface soil sampling at a nursery site

measurement taken at the beginning and end of the concrete channel (*Figure 3*). It is clear that the monthly runoff output varies seasonally, with considerably more runoff being generated during the summer months. The monthly average runoff into the channel was 8,606 M^3 in 2000, which decreased to 3,420 M^3 in 2001 and further to only 2,022 M^3 in 2002. The reduction in runoff output allowed the implementation of best management practices (BMPs) to remove sediments and sediment-adsorbed pesticides. The BMPs include polyacrylamide (PAM) delivery into the runoff stream, installation of an in-line sediment trap and a sediment settling pond, and development of a vegetative strip in the discharge channel. The arrangement of these BMPs is schematically shown in *Figure 1*. All of these BMPs work in concert to cause maximal removal of suspended solids from the runoff flow.

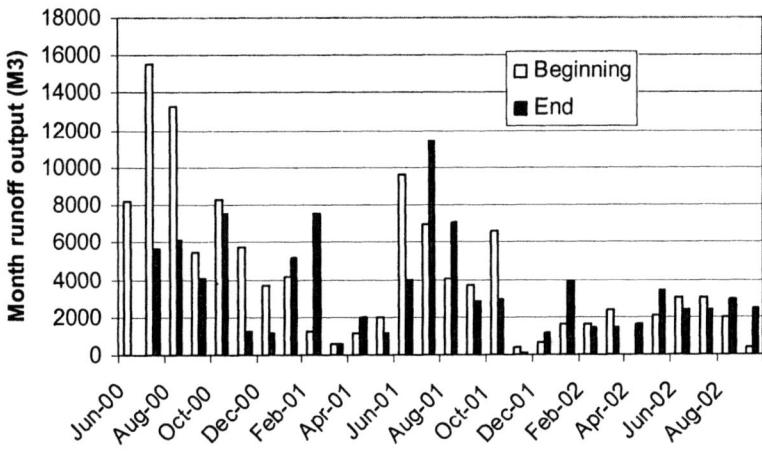

Figure 3. Monthly runoff output measured at the beginning and end of the 260-m discharge channel

Polyacrylamide (PAM)

Polyacrylamide (PAM) has been often used as a flocculent to cause aggregation and settling of suspended solids from water and to reduce soil erosion (*9, 10*). In our study, PAM solution of 10 ppm is constantly dripped into the effluent about 30 m before the sediment trap. Introduction of PAM before the sediment trap allows the polymer to be fully mixed in the runoff flow and the interaction to occur over a relatively long distance downstream.

Sediment Trap, Pond and Basins

The sediment trap and settling pond slow down the runoff flow, allowing some of the suspended solids to settle out under gravity. The sides of the sediment trap and pond may also act as a physical barrier to stop and trap suspended solids. The sediment trap is a 4 m (width) × 2 m (length) × 1 m (depth) pit, and is below the surface over which the runoff flows. The settling pond is a trianglular unlined pit, and is about 3 m deep at the center. In the discharge channel, a total of eight small basins were constructed by partially blocking the flow with sand bags. These basins are expected to further slow the runoff flow, allowing more gravitational and physical removal of suspended solids from the runoff. When significant sediment accumulation occurs in the sediment trap or pond, the sediment is mechanically cleared. When significant sediment accumulation occurs in the channel, the sediment layer is cleared manually.

Vegetative Strip

The vegetative strip was first developed in 2000 and contained canna lilies grown in plastic mesh baskets or crates submerged in water. The variety of canna lily ('Tropicanna') develops a large root system, which helps to further slow down the flow to cause settling and physical trapping of suspended solids. The canna lilies are periodically harvested for sale by the nursery grower. The canna lily strip occupies about 150 m of the 260-m long discharge channel.

Sampling and Analysis

To evaluate the effectiveness of the various BMPs, runoff samples have been collected on a monthly basis at different locations along the runoff path. The first sampling point was immediately before the PAM release point (*Figure 1*), and the last sampling point was outside of the nursery property, at about 100 m downstream from the end of the vegetative strip in the discharge channel. About 1 L of runoff water was collected by dipping the bottle opening into the center of the runoff flow. Samples were stored in an ice box and transported to the UC Riverside Campus within 4 h after sampling. Two replicates were used at each sampling point.

A 500-ml aliquot of each runoff sample was removed from the sample bottle after mixing, and passed through a 1.0-μm glass fiber membrane under vacuum. The weight of each filter paper was recorded before use. After filtration, the filter paper, along with the retained solids, was dried at 105°C for 4 h, and then weighed again. The increase in weight was used to calculate suspended solid content (mg L^{-1}) of the runoff sample.

To analyze bifenthrin and permethrin, a 250-mL aliquot of each runoff sample was transferred to a 1-L separatory funnel and extracted with 50 mL ethyl acetate by shaking for 1 min. The solvent phase was collected after phase separation. The same extraction step was repeated for two more times. The ethyl acetate extracts were combined and mixed with 50 g of anhydrous sodium sulfate to remove the residual water. The extract was then evaporated to near dryness on a rotary evaporator, and pesticide residues were redissolved in 5 ml acetone-hexane (1:1, v/v) for analysis by gas chromatograph (GC).

Quantification of bifenthrin and permethrin was carried on an Agilent 6890 GC system equipped with a micro-electron capture detector (μ-ECD). A capillary column (Agilent-5, 30 m × 0.32 mm 0.25 μm) was used with helium as the carrier gas at 2.1 mL min^{-1}. The other GC parameters were as follows: inlet temperature, 250°C; detector temperature, 300°C; oven temperature, initially 150°C for 1.0 min, ramped to 280°C at 15 °C min^{-1}, and kept at 280°C for 5 min; and injection volume, 1.0 μL. Samples were introduced in the splitless mode. Under these conditions, the retention time for bifenthrin, *cis*-permethrin, and *trans*-permethrin was 9.0, 10.2, and 10.3 min, respectively.

Results and Discussion

Sources of Pesticides in Nursery Runoff

Analysis of surface soil samples from various locations at the nursery site showed that all of the samples contained high levels of both bifenthrin and permethrin (*Table 2*). These levels were significantly greater than those found in runoff or were sometimes comparable to those found in the sediment from the settling pond or the discharge channel. Many of the soil samples contained wood chips and other potting mix materials (e.g., peat moss), indicating that potting mix had spilled on those surfaces. Samples with a high percentage of potting mix generally contained higher levels of bifenthrin and permethrin (*Table 2*). The highest pesticide levels were found at site 1, 8, 11, 13, 14, and 16. Site 1 was adjacent to an area where dead plants were composted and the soil mix recycled. The surface soil was primarily composed of potting mix, which coincided with high levels of bifenthrin and permethrin (*Table 2*). Site 8 represented a propagation area where propagation containers are filled with potting mix treated with bifenthrin. Newly propagated plants are maintained in greenhouse conditions and irrigated by misting with overhead sprinklers. Potting mix had accumulated on the surface due to spillage while filling containers. Bifenthrin concentration as high as 17,440 μg/kg was detected in these samples. Site 11 was in another outdoor area where plant containers are placed on pallets directly over a runoff channel. Runoff directly into the channel from the containers as a result of overhead irrigation and loss of potting mix was evident. Sites 13, 14, and 16 were close to the planting area where potting mix was first filled into plant containers. The samples in this area were primarily made of potting mix, which coincided with high levels of bifenthrin and permethrin residues in the soil samples (*Table 2*).

Potting mix was widespread on the soil surface throughout the nursery. It is likely that when runoff moved over the surface, it carried into the runoff some of the potting mix, along with the pesticides that were premixed into the potting mix during planting. The "hot" spots included the potting mix filling area, the dumping area, and outdoor areas where overhead sprinklers were employed for irrigation. In the potting mix filling and dumping areas, contamination of the surface with potting mix appeared to be a result of poor confinement due to spills, which was widened from traffic of tractors that were used to move plant containers between the different locations. In the outdoor areas, contamination of potting mix at the surface was apparently caused by spills under the impact of irrigation streams from overhead sprinklers. The fact that the surface was sometimes lined with relatively impermeable pads might have allowed the potting mix to be easily washed into the runoff. Accumulation of potting mix and pesticides at the other sites was likely a result of spills of potting mix from tractors carrying filled plant containers. The disturbance by tractor tires could have helped to further spread the potting mix over road surfaces throughout the nursery.

Sediment Removal by BMPs

It was observed that the upstream runoff at the nursery site consistently contained high levels of solids. *Figure 4* shows decreases of suspended solid content in runoff along the runoff path for 05/16/2000 and 06/16/2002. The suspended solid content prior to BMPs (measured before the PAM delivery point) was different between the two sampling dates, with the 05/16/2002 samples containing markedly more suspended solids than the 06/16/2002 samples. Such variations are characteristic of nursery runoff, and are a result of different activities (e.g., irrigation) occurring on different days. However, on both sampling days, it is clear that the suspended solid content rapidly decreased when the runoff traveled downstream through the sediment trap, sediment pond, and finally the vegetated channel (*Figure 4*). On both days, the greatest drop in suspended solid content occurred between the PAM delivery point and the pond, or after the sediment trap. When the PAM delivery point is used as the reference point, the suspended solid removal after the sediment trap was >90% for both sampling days (*Table 3*). More reductions further occurred in the vegetated channel. When the runoff reached the end of the vegetative strip (240 m from the pond), the overall suspended solid removal was 99.6% on 05/16/2002 and

Table 2. Concentrations of bifenthrin and permethrin found in surface soils collected from a nursery site

Sample No.	Concentration (µg/kg)			Potting mix [†]
	BF	cis-P	trans-P	
1	3,971 ± 373	4,625 ± 1,194	53 ± 13	80%
2	306 ± 4	658 ± 240	70 ± 2	20%
3	344 ± 17	1,556 ± 322	178 ± 94	5%
4	109 ± 13	471 ± 396	18 ± 5	1%
5	163 ± 18	521 ± 96	58 ± 6	<1%
6	275 ± 81	5,948 ± 651	1,478 ± 43	50%
7	307 ± 55	71 ± 124	243 ± 28	10%
8	17,440 ± 436	631 ± 831	642 ± 1,085	100%
9	163 ± 123	290 ± 138	17 ± 9	5%
10	125 ± 6	113 ± 101	23 ± 9	1%
11	2,470 ± 451	1,697 ± 369	0 ± 0	100%
12	389 ± 17	387 ± 137	24 ± 2	1%
13	1,309 ± 24	753 ± 201	21 ± 9	20%
14	852 ± 19	1,535 ± 506	16 ± 2	15%
15	430 ± 33	455 ± 89	18 ± 3	1%
16	3,055 ± 211	5,115 ± 1,669	80 ± 139	50%
17	353 ± 26	1,613 ± 765	163 ± 12	80%
18	95 ± 7	238 ± 21	27 ± 3	1%
19	303 ± 10	587 ± 184	48 ± 5	20%
20	305 ± 25	178 ± 39	35 ± 1	1%

[†] Estimated content of potting mix in soil. High potting mix content suggests spills of potting mix during movement of containers and/or irrigation.

96.9% on 06/16/2002 (*Table 3*). The suspended solid content in the runoff at the end of the vegetative strip was only 15 and 32 mg L^{-1} on 05/16/2002 and 06/16/2002, respectively. The reduction in total mass of suspended solids in runoff was greater than indicated by the sediment concentration data above. The total mass of suspended solids in runoff is equal to the product of runoff volume and suspended solid concentration. Consequently the decrease in runoff over the last few years (Figure 3) has further reduced sediment movement off-site relative to years before irrigation and pesticide BMPs were implemented.

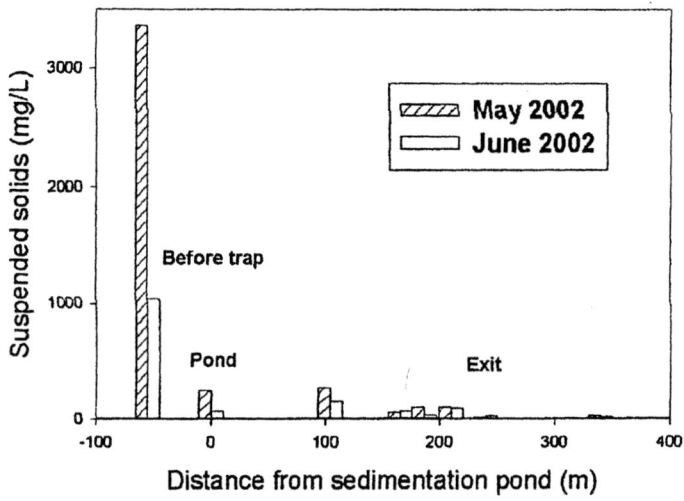

Figure 4. Reduction in suspended solid content of runoff at different locations along the runoff path

In the runoff path, sediment aggregation, disaggregation, settling and resuspension constantly occur. The importance of each of these process changes momentarily as a function of particle size and density, the flow velocity, the depth of the water column, and geometrical configuration of the runoff path (*11*). Under gravitational force, large and dense particles tend to settle to the sediment layer more rapidly and hence closer to the point of origination (e.g., in the vicinity of sediment trap and pond). Fine particles will travel over a longer distance downstream. The settled sediment particles may be resuspended due to currents and other activities. Settling rate will increase if the flow speed decreases. All these factors may have contributed to the removal of suspended solids from the runoff at the nursery site. In addition, PAM is known to enhance aggregation of suspended solids, further accelerating the gravitational settling of suspended solids from the moving water. The settling of suspended solids could be further enhanced when the movement of suspended particles was stopped by a physical barrier, such as by the sides of sediment trap or pond, sand bags in the basins along the discharge channel, and extended plant roots in the vegetative strip.

Table 3. Percentages of reduction in sediment level in runoff water along the runoff path (% of level measured before the PAM delivery point)

Position	May 2002		June 2002	
	Concentration (mg L^{-1})	Removal (%)	Concentration (mg L^{-1})	Removal (%)
Before PAM	3360	-	1404	-
Pond	254	-92.4	70	-93.3
104 m†	276	-91.8	159	-84.8
166 m	60	-98.2	66	-93.7
187 m	102	-97.0	34	-96.7
210 m	101	-97.0	98	-90.6
240 m‡	15	-99.6	32	-96.9
340 m	26	-99.2	23	-97.8

†Beginning of vegetative strip.
‡End of vegetative strip.

Pesticide Removal by BMPs

Runoff samples were also analyzed for bifenthrin and permethrin. Bifenthrin concentrations in runoff generally decreased as the runoff moved through the sediment trap, pond, and the discharge channel (*Figure 5*). For instance, on 05/16/2002, the initial bifenthrin level in runoff before the PAM release point was 10.6 µg L^{-1}, which decreased to 0.87 µg L^{-1} at the end of the vegetative strip. On 06/16/2002, the initial bifenthrin level was 3.2 µg L^{-1}, which decreased to

0.28 µg L^{-1} at the vegetative strip. Using the concentration before the PAM release point as the reference point, the reduction in bifenthrin concentration in the runoff was 91.8% on 05/16/2002, and 91.3% on 06/16/2002 (*Table 4*). The greatest decrease occurred after the sediment trap, but further decreases occurred through the vegetated channel. The pesticide removal was apparently correlated with the removal of suspended solids caused by the various BMPs along the runoff path.

Table 4. Reductions in bifenthrin level in runoff along the runoff path (% of level measured before the PAM delivery point)

Position	May 2002		June 2002	
	Concentration (ppb)	Removal (%)	Concentration (ppb)	Removal (%)
Before PAM	10.56	-	3.18	-
Pond	1.41	-86.7	0.93	-70.7
104 m†	9.27	-12.2	1.11	-65.0
166 m	4.26	-59.6	0.55	-82.8
187 m	2.83	-73.2	0.43	-86.6
210 m	1.68	-84.1	0.95	-70.2
240 m‡	0.87	-91.8	0.28	-91.3
340 m	0.96	-90.9	0.30	-90.7

†Beginning of vegetative strip.
‡End of vegetative strip.

The decrease in the concentration of *cis*- and *trans*-permethrin along the runoff path was similar to that of bifenthrin (*Figure 6*). The greatest decrease occurred after the sediment trap, but further reduction occurred also in the vegetated channel. On 05/16/2002, the initial *cis*-permethrin level was 24.5 µg L^{-1}, which decreased to 1.2 µg L^{-1} at the end of the vegetative strip. On 06/16/2002, the initial *cis*-permethrin level was 2.4 µg L^{-1}, which decreased to below the detection limit (0.05µg L^{-1}) at the end of the vegetative strip (*Figure 6*). Similar decreases were also recorded for the *trans* isomer of permethrin. The reduction at the end of the vegetative channel was equivalent to 95.2 and 100% removal for *cis*- and *trans*-permethrin, respectively, for the day of 05/16/2002 (*Table 5*). The respective removal rates were 94.1% and 100% for the day of

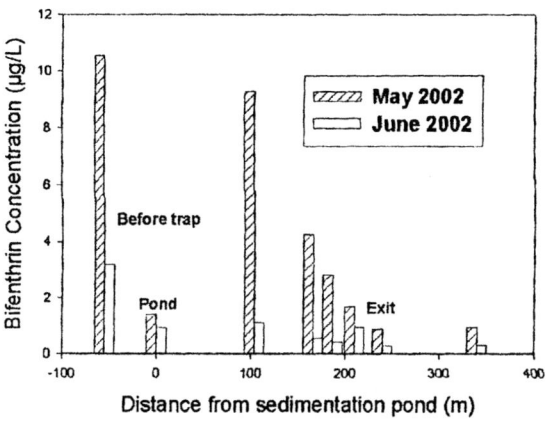

Figure 5. Reduction in bifenthrin in runoff at different locations

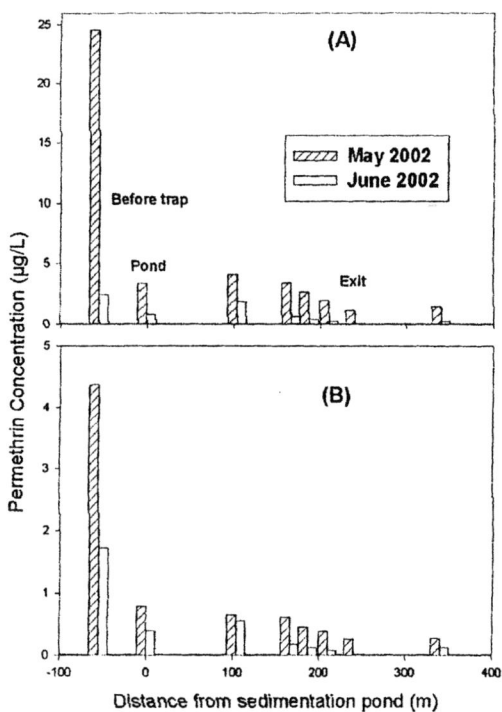

Figure 6. Reduction in permethrin concentration in runoff at different locations. (A) cis-permethrin; (B) trans-permethrin.

Table 5. Reductions in permethrin level in runoff along the runoff path (% of level measured before the PAM delivery point)

Position	cis-Permethrin			
	May 2002		June 2002	
	Concentration (ppb)	Removal (%)	Concentration (ppb)	Removal (%)
Before PAM	24.55	-	2.44	-
Pond	3.34	-86.4	0.85	-81.9
104 m†	.4.08	-83.4	1.87	-85.1
166 m	3.41	-86.1	0.62	-86.0
187 m	2.68	-89.1	0.36	-89.7
210 m	1.96	-92.0	0.24	-91.3
240 m‡	1.19	-95.2	ND	-94.1
340 m	1.45	-94.1	0.22	-93.8
	trans-Permethrin			
Before PAM	4.37	-	1.73	-
Pond	0.79	-65.1	0.39	-77.3
104 m†	0.65	-23.4	0.55	-68.1
166 m	0.61	-74.7	0.17	-90.2
187 m	0.45	-85.2	.0.12	-92.9
210 m	0.38	-90.1	0.07	-95.7
240 m‡	0.26	-100	ND	-100
340 m	0.27	-90.8	0.13	-92.6

†Beginning of vegetative strip.
‡End of vegetative strip.

06/16/2002 (*Table 5*). Again, the removal of permethrin appeared to correlate closely with that of suspended solids in the runoff.

It is evident that although the overall decreasing trend of bifenthrin or permethrin concentration was similar to that of suspended solid content in the runoff, the removal ratio was not directly proportional. At any given location, the removal of suspended solids was consistently greater than the removal of pesticides. This may be attributed to the fact that settling of suspended solids over distance is a particle size discretion process. Most large particles may have settled out in the front section, which coincided with a steep decrease in suspended solid content that was measured using a mass unit, i.e., mg L^{-1} (*Figure 4*). However, large particles tend to have smaller specific surface areas and lower organic matter or clay content than small particles. Therefore, less pesticides would be adsorbed to the particles settled out at the beginning of the runoff path than to the small particles that traveled over some distance downstream. It can be further speculated that in order to further reduce pesticide load in the runoff, it would be essential to eliminate the small particles in the runoff flow.

Conclusions

Uncontrolled runoff at commercial nurseries can result in pesticide and nutrient contamination of waterbodies. We evaluated the use of several best management practices (BMPs) to reduce pesticide load in runoff at a nursery site in southern California. The BMPs include sediment trap, sediment pond, application of polyacrylamide (PAM), sediment basins, and vegetative filter strip. All of these BMPs operated on the principle of physical removal of suspended solids in the runoff. These practices were found to be effective in removing suspended solids, as well as strong-adsorbing pesticides such as bifenthrin and permethrin, from the runoff. Overall, the removal of suspended solids by these BMPs approached 97-99%, and the reduction of bifenthrin and permethrin was consistently >90%. These BMPs are of low cost and low maintenance, and may be easily adopted by other nursery growers or other end users. Our study also suggests that to completely remove bifenthrin or permethrin from the runoff, it may be necessary to also remove the very fine suspended solids in the runoff. The synthetic pyrethroids are detrimental to aquatic organisms when present in the aquatic systems at very low concentration. However, the fate and distribution of synthetic pyrethroids in and after nursery runoff is still poorly understood. The interaction of suspended solids with the transport and toxicity of synthetic pyrethroids needs to be further investigated to address the on-going regulatory challenges.

Acknowledgements

Support of this work was provided by the California Department of Pesticide Regulation, the California Department of Food and Agriculture Fertilizer Research and Education Program, the Orange County Farm Bureau, State Water Resources Control Board/Santa Ana Regional Water Quality Control Board EPA 319(h).

Reference:

1. Walters, J.; Kim, D.; Goh, K. Preliminary results of pesticide analysis of monthly surface water monitoring for the red and imported fire ant project in Orange County, Department of Pesticide Regulations, Sacramento, CA 92815; http://www.cdpr.ca.gov/docs/rifa/reports.htm.
2. USEPA, Newport Bay Toxics TMDLs, Part C, Organophosphate (OP) Pesticides; http://www.forester.net/sw.html.
3. Werner, I.; Deanovic, L.A.; Hinton, D.E.; Henderson, J.D.; de Oliveira, G.H.; Wilson, B.W.; Krueger, W.; Wallender, W.W.; Oliver, M.N.; Zalom, F.G. Toxicity of stormwater runoff after dormant spray application of diazinon and esfenvalerate (Asana) in a French prune orchard, Glenn County, California, USA. *Environ. Contam. Toxicol.* **2002**, *68*, 29-36.
4. Domagalski, J.L.; Dubrovsky, N.M.; Kratzer, C.R. Pesticides in the San Joaquin River, California: Inputs from dormant sprayed orchards. *J. Environ. Qual.* **1997**, *26*, 454-465.
5. Ma, Q.L.; Smith, A.E.; Hook, J.E.; Smith, R.E.; Bridges, D.C. Water runoff and pesticide transport from a golf course fairway: Observations vs. Opus model simulations. *J. Environ. Qual.* **1999**, *28*, 1463-1473.
6. Domagalski, J. Pesticides and pesticide degradation products in stormwater runoff: Sacramento River Basin, California. *Water Res. Bull.* **1996**, *32*, 953-964.
7. Kuivila, K.M.; Foe, C.G. Concentrations, transport and biological effects of dormant spray pesticides in the San Francisco Estuary, California. *Environ. Toxicol. Chem.* **1995**, *14*, 1141-1150.
8. Gan, J. PesticideWise databases; http://www.pw.ucr.edu/.
9. Lentz, R.D.; Shainberg, I.; Sojka, R.E.; Carter, D.L. Preventing irrigation-induced furrow erosion with small applications of polymers. *Soil Sci. Soc. Am. J.* **1992**, *56*, 1926-1932.
10. Lentz, R.D.; Sojka, R.E. Field results using polyacrylamides to manage furrow erosion and infiltration. *Soil Sci.* **1994**, *158*, 274-283.
11. Lick, W.; The transport of sediments in aquatic systems. In Dickson K. et al. (Eds.) *Fate and Effects of Sediment-Bound Chemicals in Aquatic Systems.* Pergamon Press: New York, USA, 1987; pp. 61-74.

Chapter 17

Impacts of Surfactant Adjuvants on Pesticide Availability and Transport in Soils

Kurt D. Pennell[1], Ahmet Karagunduz[2], and Michael H. Young[3]

[1]School of Civil and Environmental Engineering, Georgia Institute of Technology, Atlanta, GA 30332-0512
[2]Department of Environmental Engineering, Gebze Institute of Technology, Gebze, Turkey
[3]Division of Hydrologic Sciences, Desert Research Institute, 755 East Flamingo Road, Las Vegas, NV 89119

Surfactants are frequently added to pesticide and herbicide formulations as adjuvants to improve handling, delivery and effectiveness. From a regulatory perspective such additives are generally considered to be inert, and their influence on co-contaminant fate and transport processes has been largely ignored. The objective of this chapter is to illustrate the potential effects of representative surfactant adjuvants on the phase distribution and availability of hydrophobic organic compounds (HOCs), soil water retention and water flow in unsaturated soils. Although the addition of surfactants at concentrations above the critical micelle concentration (CMC) is shown to enhance the total aqueous-phase concentration of HOCs, the free (non-micellar) aqueous phase HOC concentration decreases with increasing surfactant concentration. Results of pressure-saturation studies and one-dimensional column experiments demonstrate that surfactants can substantially reduce soil water retention and alter unsaturated water flow. These findings demonstrate the need to carefully consider the influence of surfactant adjuvants on both soil water characteristics and agrochemical fate and transport in the environment.

Introduction

Herbicides are widely used to control undesirable plant growth in agricultural systems, accounting for approximately 65% of the pesticide market in the United States *(1)*. Almost all herbicides require some type of adjuvant in the formulated product to improve handling. In addition, adjuvants may be added to spray tank solutions to improve delivery and effectiveness of herbicides. A number of adjuvants have been utilized in pesticide formulations, including surfactants, solvents, preservatives, thickeners, carriers, and antifreeze agents. Of these, surfactants comprise approximately 45% of the U.S. market, equivalent to approximately 400 million pounds of surfactant applied in 1997 *(2)*. Surfactant adjuvants typically constitute 5 to 10% (50,000 to 100,000 mg/L) of the formulated herbicide product and approximately 0.1 to 0.5% (1,000 to 5,000 mg/L) of the spray tank solution. Thus, the amount of surfactant applied to agricultural crops can be significant, particularly if repeated applications are required throughout the growing season. This chapter addresses the potential impacts of surfactant adjuvants on the fate and transport of agrochemicals in soils. The influence of surfactants on the aqueous solubility, phase distribution and availability of hydrophobic organic compounds (HOCs) will be discussed in detail. In addition, the effects of surfactant addition on soil water retention, unsaturated water flow and solute transport will be presented.

Surfactant Properties

Surfactants are amphiphilic compounds, possessing both hydrophobic (nonpolar) and hydrophilic (polar) moieties. The hydrophobic "tail" group typically consists of a long chain (e.g., C8-C20) hydrocarbon, whereas the hydrophilic "head" group is either an ionic or highly ethoxylated species. This unique chemical structure results in a strong tendency for surfactants to accumulate at phase interfaces, including the gas-liquid and solid-liquid interfaces. For example, the addition of even small amounts of surfactant to water markedly reduces the interfacial tension (surface tension) of the air-water interface. This characteristic behavior serves as the basis for the term surfactant, which is a contraction of the phrase "surface active agent".

Thousands of surfactants are produced commercially worldwide, with applications ranging from pharmaceuticals and food additives to soaps and detergents. Surfactants are typically classified as either anionic, cationic, nonionic or amphoteric depending upon the nature of their hydrophilic head group. Nonionic surfactants are widely used in agricultural applications because of their relatively low cost, generally low toxicity, and tolerance to varying solution conditions (e.g., electrolyte concentration). The most common nonionic surfactants used as adjuvants include tallow amine ethoxylate, nonylphenol and octylphenol ethoxylates, alcohol ethoxylates, block polymers, and castor oil ethoxylates *(2)*.

The hydrophilic head group of nonionic surfactants contains either hydroxyl groups, or more commonly, ethylene oxide (EO) polymer chains. The relative polarity of nonionic surfactants is often characterized by the hydrophile-lipophile balance (HLB), defined as *(3)*:

$$HLB = 20 X \frac{MW_H}{MW_H + MW_L} = \frac{\% \, wt. \, EO}{5} \qquad (1)$$

where *MW* is molecular weight, *EO* is ethylene oxide, and the subscripts *H* and *L* refer to the hydropholic and lipophilic groups, respectively. The HLB scale (0 to ~20) is used to classify nonionic surfactants with respect to their behavior in water and commercial application. Surfactants with low HLB numbers (3-6) are poorly dispersed in water, and used primarily as water-in-oil emulsifiers. In contrast, nonionic surfactants with HLB numbers greater than 13 are often used as detergents and tend to form clear solutions when dissolved in water. Examples of representative nonionic surfactant adjuvants are provided in Table 1.

Table 1. **Properties of representative nonionic surfactants used as pesticide adjuvants** *(3)*.

Surfactant Name	Molecular Formula	Molecular Weight (g/mole)	HLB	CMC (mg/L)
Dodecyl alcohol ethoxylate (n=9)	$C_{12}H_{25}O(C_2H_4O)_9H$	583	13.6	50-65
Octylphenol ethoxylate (n=9.5)	$C_8H_{19}C_6H_4O(C_2H_4O)_{9.5}H$	625	13.5	110-150
Polyoxythylene sorbitan monooleate (n=20)	$C_{18}H_{34}O_2C_6H_{10}O_4(C_2H_4O)_{20}$	1310	15.0	35-45

HLB = hydrophile-lipophile balance, CMC = critical micelle concentration.

Micellar Solubilization

When surfactants are added to water at low concentrations, surfactant molecules exist as single monomers in solution. As the surfactant concentration is increased, the hydrophobic moieties of individual surfactant monomers associate with one another to form micelles. The surfactant concentration corresponding to the onset of micelle formation is referred to as the critical micelle concentration or CMC. The CMC of most nonionic surfactants falls within the range of 50 to 200 mg/L, and thus, is exceeded in most agrochemical applications. The characteristic number of monomers in each micelle is referred to as the aggregation number (N_A), and remains essentially constant until the solubility limit of the surfactant is approached. Therefore, as the surfactant concentration is increased above the CMC, the number of surfactant micelles increases, while the number of monomers in solution remains constant,. As shown in Figure 1, the mass fraction of surfactant existing in micelles increases as more surfactant is added to solution, while the fraction of surfactant existing as monomers decreases (Figure 1).

In most aqueous systems, surfactant micelles consist of a hydrophobic core surrounded by a hydrophilic mantle, which is widely referred to as a Winsor

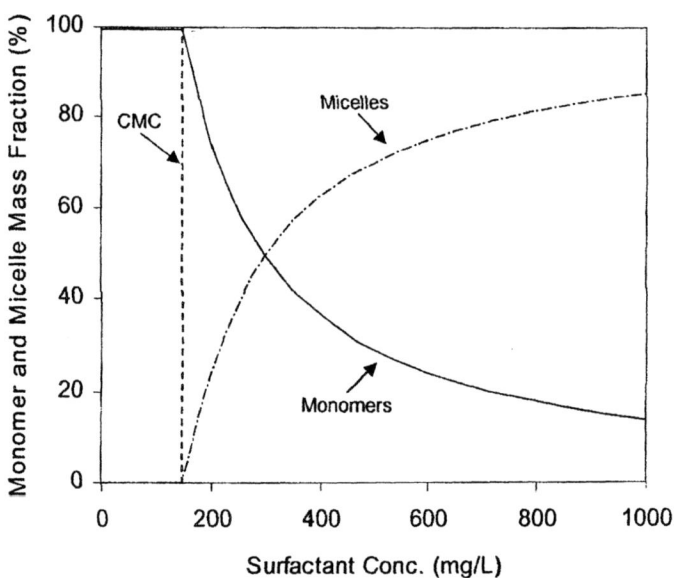

Figure 1. Change in surfactant monomer and micelle mass fraction as a function of total surfactant concentration.

Type I microemulsion *(3)*. When a Type I surfactant microemulsion is contacted with excess HOC, a linear enhancement in HOC solubility is typically observed for surfactant concentration above the CMC. Two examples of this behavior are provided in Figure 2, for 1,1-bis(p-chlorophenyl)-2,2,2-trichloroehtane (DDT) and hexachlorobenzene (HCB) in the presence of Triton X-100 (an octylphenol ethoxylate). The linear enhancement in solubility above the CMC is attributed to the incorporation or partitioning of HOCs into the hydrophobic core of surfactant micelles *(3-5)*. Below the CMC, a slight increase in solubility is observed for both DDT and HCB, although this effect is generally considered to be minimal for more water soluble HOCs *(4)*.

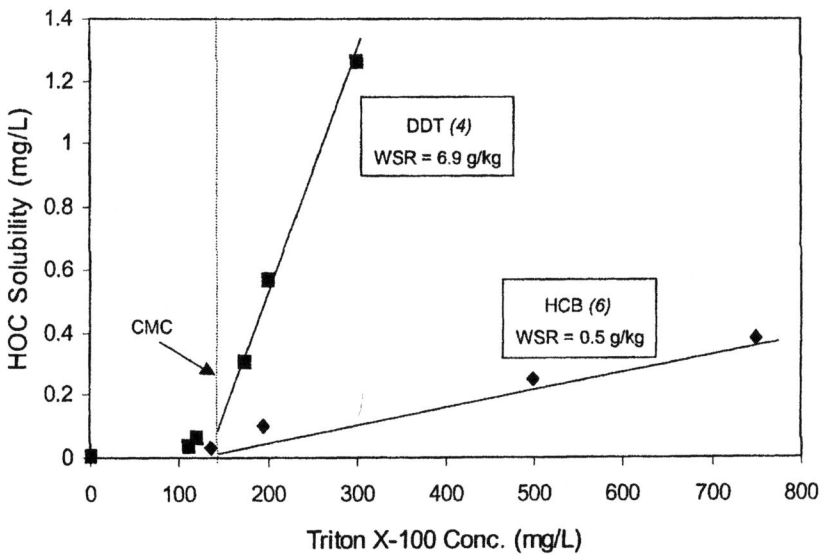

Figure 2. Aqueous solubility of DDT (4) and HCB (6) as a function of Triton X-100 concentration.

The capacity of a surfactant to solubilize a particular HOC can be described by the weight solubilization ratio (WSR), defined as the mass of HOC solubilized per mass of surfactant in micellar form. The WSR corresponds to the slope of the linear portion of the solubility curve above the CMC, and may be expressed as:

$$WSR = \frac{C_w^{HOC} - C_{w,CMC}^{HOC}}{C_w^{surf} - C_{w,CMC}^{surf}} \qquad (2)$$

where C_w^{HOC} is the total concentration of HOC in solution (mg/L), $C_{w,CMC}^{HOC}$ is the concentration of HOC in solution at the CMC of the surfactant (mg/L), C_w^{surf} is the total concentration of surfactant in solution (g/L), $C_{w,CMC}^{surf}$ is the CMC of the surfactant (g/L). The effects of surfactants on HOC solubility may also be described in terms of the HOC distribution between the aqueous phase and surfactant monomers and micelles. Based on this conceptual approach, the apparent aqueous solubility of an HOC in a surfactant solution may be written as (5):

$$\frac{C_{w,sol}^{HOC*}}{C_{w,sol}^{HOC}} = 1 + C_{w,mn}^{surf} K_{mn} + C_{w,mc}^{surf} K_{mc} \qquad (3)$$

where $C_{w,sol}^{HOC*}$ is the apparent or observed solubility of the HOC (mg/L), $C_{w,sol}^{HOC}$ is the solubility of the HOC in water (mg/L), $C_{w,mn}^{surf}$ is the concentration of surfactant monomers in solution (g/L), K_{mn} is the monomer-water partition coefficient (L/g), $C_{w,mc}^{surf}$ is the concentration of surfactant in micellar form (g/L), and K_{mc} is the micelle-water partition coefficient (L/g). Several correlations have been developed to relate measured K_{mc} values to both HOC properties (e.g., octanol-water partition coefficient, K_{ow}) and surfactant properties (e.g., N_A, HLB) (7-9). In general, the magnitude of the observed solubility enhancement due to micellar solubilization is more pronounced for strongly hydrophobic compounds (i.e., higher K_{ow}). Nevertheless, the total mass of HOC in solution is usually greater for hydrophobic compounds with relatively low K_{ow} values (i.e., more hydrophilic). The latter point is important to recognize if the intent of surfactant addition is to increase herbicide mass delivery.

HOC Phase Distribution in Surfactant-Soil Systems

The previous discussion was limited to systems containing surfactant, HOC and water. The presence of soil substantially increases the level of complexity because both the surfactant and HOC may interact with the solid phase via sorption mechanisms. Furthermore, HOCs may interact with the adsorbed surfactant phase, thereby increasing the apparent HOC sorption capacity of the solid phase. Upon addition of surfactant to a soil-water system, surfactant monomers will readily adsorb at the gas-liquid and solid-liquid interfaces. In general, the concentration of surfactant in the aqueous phase remains low until these interfaces are completely saturated with monomers. For this reason, the surface tension of water decreases, and the amount of surfactant adsorbed by the solid phase increases, until the CMC is reached. Above the CMC, both of these properties remain relatively constant. A schematic representation of a three-phase system (gas-liquid-solid), containing soil, water, surfactant and HOC is shown in Figure 3.

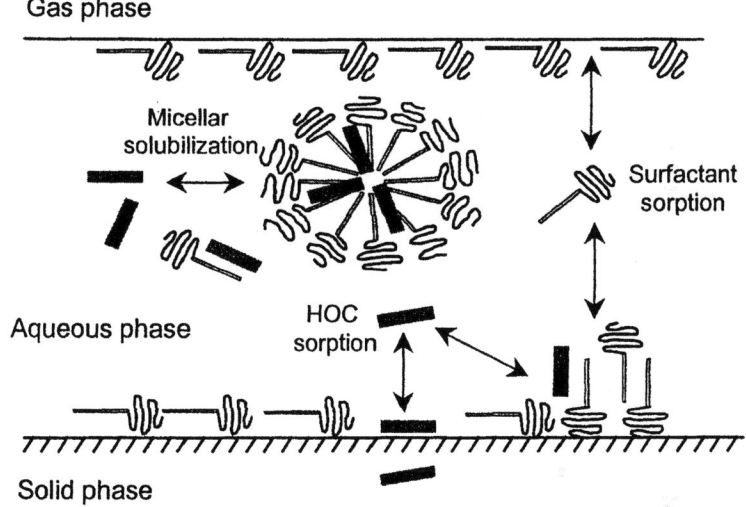

Figure 3. Schematic diagram of a three-phase soil system (gas-liquid-solid) containing surfactant and HOC.

Surfactant Sorption

The sorption of surfactants by soils and other solids is typically described using the Langmuir equation (*3*):

$$C_s^{surf} = \frac{C_{s,\max}^{surf} b C_w^{surf}}{1 + b C_w^{surf}} \qquad (4)$$

where C_s^{surf} is the solid-phase concentration of surfactant (g/kg), $C_{s,max}^{surf}$ is the maximum or limiting solid-phase concentration of surfactant (g/kg), C_w^{surf} is the aqueous-phase concentration of surfactant (g/L), and b represents the ratio of the adsorption and desorption rate constants (L/g). Due to their amphiphilic properties, surfactants may be sorbed by solid phases as a result of both hydrophilic and hydrophobic interactions. At low surface coverages, surfactant monomers are often considered to be oriented parallel to mineral surfaces, forming a monolayer as the CMC is approached (10). When the surfactant concentration increases above the CMC, the hydrophobic groups may be displaced from the surface. These hydrophobic groups will then be oriented toward the aqueous phase and may interact with the hydrophobic groups of other surfactant monomers, resulting in bilayer formation (Figure 3).

The maximum sorption capacity of nonionic surfactants typically ranges from approximately 0.1 to 1.5 g/kg for clean sands, and up to 10.0 g/kg for natural soils containing low levels of organic carbon. Representative sorption data for Tween 80 and Triton X-100 are provided in Table 2. Based on estimates of the cross-sectional area of surfactant molecules, the maximum sorption capacity corresponds to surface coverages ranging from 0.5 to 5.0 monolayers (6, 11).

Table 2. Langmuir sorption parameters for selected surfactant-soil systems.

Surfactant	Soil	$C_{s,max}^{surf}$ (g/kg)	b (L/g)	Ref.
Triton X-100	Appling sandy loam (TOC = 0.68%, SSA = 4.4 m²/g)	1.18	3.0	(6)
Triton X-100	Lincoln fine sand (TOC=0.05%, SSA=3.0 m²/g)	4.11	1.6	(11)
Triton X-100	F-70 Ottawa sand (TOC <0.01%, SSA = 0.15 m²/g)	0.06	9.0	(6)
Tween 80	Appling sandy loam (TOC = 0.68%, SSA = 4.4 m²/g)	6.00	8.0	(6)
Tween 80	F-70 Ottawa sand (TOC <0.01%, SSA = 0.15 m²/g)	0.16	29.0	(12)

TOC = total organic carbon content, SSA = specific surface area.

Coupled HOC and Surfactant Sorption

Based on the discussion above, it is apparent that the magnitude of surfactant sorption can be substantial, even if the resulting isotherm approaches a limiting value at or near CMC. In the presence of an adsorbed surfactant phase and the absence of surfactant micelles in solution, sorption of HOCs has been found to increase substantially (*13-15*). Such behavior is attributed to interactions between HOCs and the hydrophobic moieties of the adsorbed surfactant phase. For surfactant concentrations greater than the CMC, however, further addition of surfactant has been shown to incrementally reduce HOC sorption by the solid phase (*6,15,16*).

The influence of surfactant sorption on the phase distribution of HOCs has been investigated by a number of researchers (e.g., *13-16*). As an example, consider the sorption of Tween 80 by Appling soil. At the maximum sorption capacity of 6.0 g/kg (Table 2), the organic carbon content of the soil would increase by nearly 60%, from 0.68% (wt.) to approximately 1.08% (wt.). Thus, surfactant sorption has the potential to impart sizable increases in the effective organic carbon content of soils. The resulting effect of surfactant sorption on the distribution of HCB between the aqueous and solid phases, over a concentration range of 0 to 840 mg/L, is illustrated in Figure 4 (*17*). In the absence of Tween 80, the equilibrium solid-phase distribution coefficient (K_D) of HCB is 527 L/kg, which corresponds to a Log K_{oc} value of 4.9. As the aqueous phase concentration of surfactant is increased from 50 to 850 mg/L, the corresponding HCB distribution coefficient decreases incrementally, from 76 to 10 L/kg. The observed decline in the HCB distribution coefficient is attributed to the greater tendency for HCB to be incorporated within the aqueous micellar phase relative to sorption by either the solid phase or the adsorbed surfactant phase. At surfactant concentrations below the CMC (i.e., < 35 mg/L), the HCB distribution coefficient would be expected to increase as a result of HCB sorption by the adsorbed surfactant phase, as reported by Sun et al. (*16*). Such behavior is particularly apparent for strongly hydrophobic compounds (high K_{ow}), and for soils or soil constituents containing negligible amounts of native organic matter (*13*).

To describe the effects of coupled sorption processes on the phase distribution of HOCs in surfactant-soil systems, an apparent soil-water distribution coefficient (K_D^*) can be developed from Eq. 3 (*14*):

$$K_D^* = \frac{K_D \left(1 + C_s^{surf/oc} K_D^{surf/oc}\right)}{\left(1 + C_{w,mn}^{surf} K_{mn} + C_{w,mc}^{surf} K_{mc}\right)} \quad (5)$$

where $C_s^{surf/oc}$ is the ratio of the sorbed surfactant concentration (g/kg) and the native organic carbon concentrations (g/kg), and $K_D^{surf/oc}$ is the ratio of the sorbed surfactant HOC distribution coefficient (L/g) and organic carbon HOC distribution coefficient (L/g). Based on this approach, the mass distribution of HCB among the solid phase, aqueous phase (free water), monomers and micelles can be computed (Figure 5). These data clearly illustrate that as the aqueous phase concentration of Tween-80 increases from 0 to 1,000 mg/L, the mass fraction of HCB associated with the solid phase decreases, whereas the fraction of HCB associated with surfactant micelles increases. In fact, the mass fraction of HCB present in the aqueous phase approaches 80% at a Tween 80 concentration of 1,000 mg/L. However, the mass fraction of HCB in the free aqueous phase (non-micellar) decreases to less than 1%. This corresponds to a reduction in the free aqueous phase concentration of HCB from 7 μg/L in the to 1.3 μg/L. If one assumes that HOCs must exist in the free aqueous phase in order to be bioavailable, these results that surfactant addition may actually inhibit direct utilization of HOCs by microbial populations. This analysis serves to emphasize the importance of carefully quantifying the effects of surfactants on the phase distribution of HOCs in soil systems. In addition, these observations may explain, in part, observed reductions in biological activity toward HOCs in the presence of otherwise compatible surfactants (*18*).

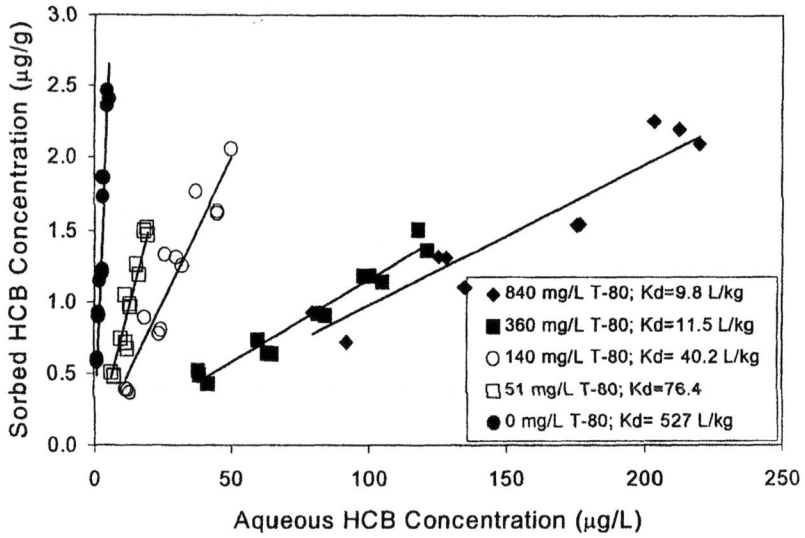

Figure 4. Effect of Tween 80 on the sorption of HCB by Appling soil (17).

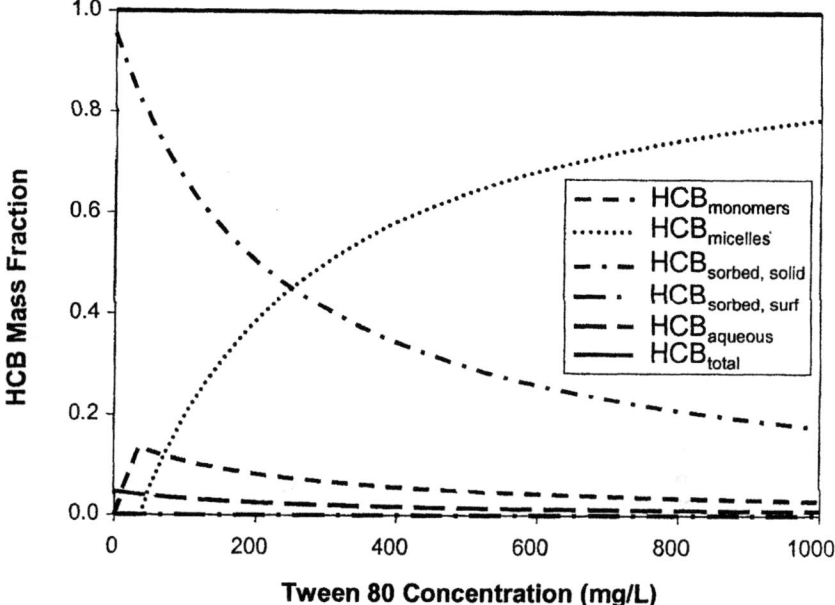

Figure 5. Phase distribution of HCB in a Tween 80-Appling soil system containing 3.0 µg of HCB, 1.0 g soil, and 20 mL of water. Input parameters: TOC = 6.8 g/kg; CMC = 35 mg/L; b = 8.0 L/g; $C_{s,max}^{surf}$ = 5.99 g/kg; K_D = 0.41 L/g; $K_D^{surf/oc}$ = 0.0 L/g; K_{mn} = 95.9 L/g; K_{mc} = 95.9 L/g.

Effects of Surfactants on Soil Water Retention and Flow

In addition to altering the phase distribution of HOCs in soil systems, surfactants may influence soil hydraulic properties due to changes in both surface tension of water and the angle of fluid contact with the solid phase. These phenomena can lead to increased infiltration depth during pesticide application and the creation of positive flow gradients from regions containing surfactant to regions free of surfactant. Demond et al. (*19*) investigated the effects of a cationic surfactant, cetyltrimethylammonium bromide (CTAB), on soil water pressure-saturation relationships. Observed changes in pressure-saturation relationships were described by accounting for concentration-dependent changes in surface tension and contact angle. In the presence of a nonionic surfactant, Triton X-100, soil water retention by F-70 Ottawa sand was shown to decrease incrementally as the surfactant concentration increased up to the CMC (20). For example, at a negative pressure head of 20 cm, the volumetric water content decreased from an initial value (no surfactant) of 0.22 mL/mL to 0.12 mL/mL at surfactant concentration of 750 mg/L. At Triton X-100 concentrations greater than the CMC, no additional changes in soil water retention were observed.

Soil water retention (pressure-saturation) relationships are frequently described using the van Genuchten (VG) equation (21),

$$\frac{\theta - \theta_r}{\theta_s - \theta_r} = \frac{1}{\left(1 + |\alpha h|^n\right)^{(1-1/n)}} \tag{6}$$

where θ is the volumetric water content (mL/mL), with the subscripts s and r refer to saturated and residual water contents, respectively, h is the negative pressure head or suction (m), and α and n are fitting parameters. To account for the effects surfactant-induced changes on surface tension and contact angle, the following scaling factor was developed (20):

$$h_0 = \left(\frac{\gamma_0 \cos\omega_0}{\gamma_1 \cos\omega_0}\right) h_1 \tag{7}$$

where γ is the surface tension of water (g/s^2), ω is the contact angle between the liquid and solid phases, and the subscripts 0 and 1 refer to pure water or a reference solution and an aqueous surfactant solution, respectively. The ability of the scaling factor (Eq. 6) to account for the effects of Triton X-100 on soil water retention curves is illustrated in Figure 6. Here, predictions based on the unscaled (VG parameters obtained in the absence of surfactant) and scaled (curves adjusted using equation 6) VG equation are compared to soil water content data obtained in the presence of 750 mg/L Triton X-100. The unscaled VG relationship greatly over predicted soil water content, whereas the inclusion of Eq. 6, using independently measured surface tension and contact angle data, provided an accurate description of the soil water retention relationship. The utility of a simplified scaling factor that employs a single parameter, β, is also shown in Figure 6 (20).

In a related set of experiments, one-dimensional soil column experiments were conducted to assess the impact of surfactants on water flow and coupled

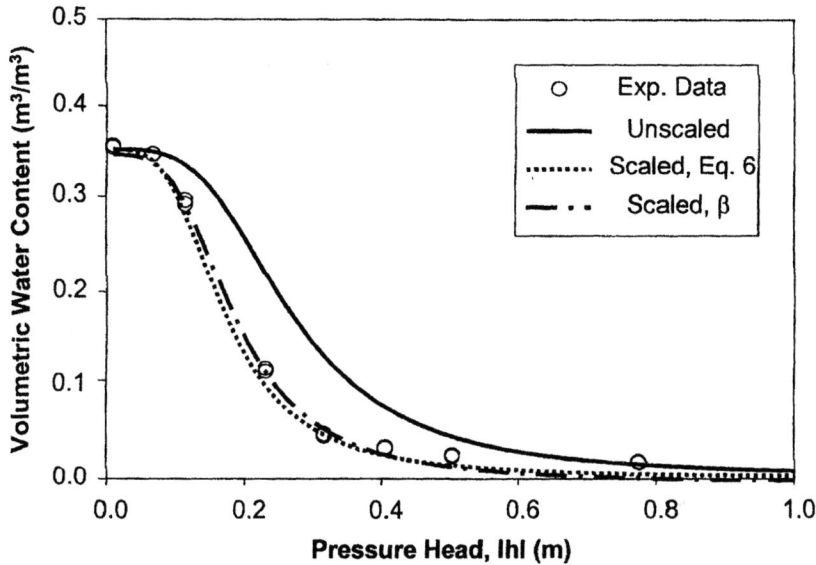

Figure 6. Measured and fitted soil water retention curves for F-70 Ottawa sand at 750 mg/L of Triton X-100.

transport of a non-reactive solute (6). Substantial reductions in soil water pressure were observed as the surfactant pulse traveled through the unsaturated soil columns. The change in water pressure resulting in a sharp reduction in soil water content, and a corresponding increase in water flux (drainage) from the column. Measured changes in soil water pressure, recorded by tensiometers located 5 and 25 cm from the column inlet, are shown in Figure 7 for a representative soil column with an initial water content of 0.195 mL/mL. Transport of a non-reactive solute, introduced with the surfactant pulse, was accelerated due to the increased water drainage described above. These results suggest that water infiltration and solute transport will be enhanced in the presence of surfactants. However, water infiltration events following a surfactant application may be reduced initially due to the lower antecedent soil water content arising from surfactant-induced water drainage.

Conclusions

The results presented herein clearly demonstrate that surfactant adjuvants have the potential to impart substantial changes in aqueous solubility and phase

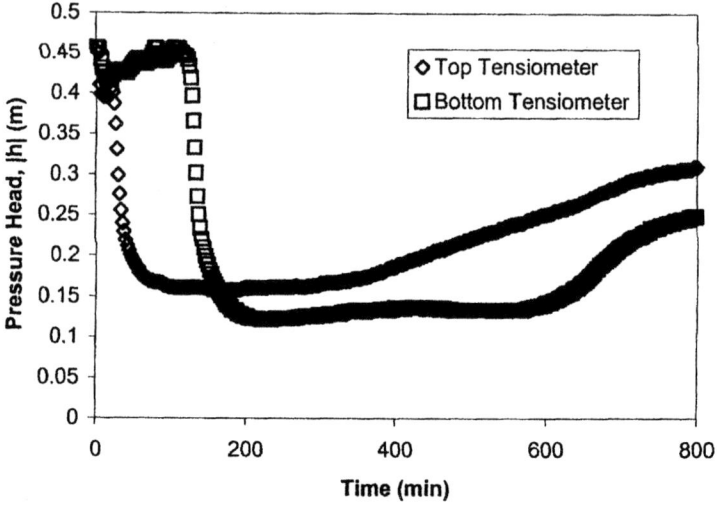

Figure 7. Influence of Triton X-100 pulse injection on the soil water pressure in an unsaturated column packed with F-70 Ottawa sand.

distribution of HOCs in soil systems. The magnitude of these effects will be most evident for strongly hydrophobic compounds, which exhibit the greatest affinity for surfactant micelles and adsorbed surfactant phases. In addition, surfactant-induced changes in the surface tension of water and angle of fluid contact with the solid phase can substantially alter soil water retention and unsaturated water flow. As a result, both infiltration depth and solute transport are likely to increase in the presence of surfactants. In closing, surfactant adjuvants should not be viewed as "inert" ingredients in pesticide formulations, and their potential impacts should be accounted for to accurately interpret and predict agrochemical fate and transport in the environment.

Acknowledgements

This work was supported by the Herty Foundation, Research and Development Center, Traditional Paper Industries Program in Pulp and Paper and the by the U.S. Environmental Protection Agency, National Center for Environmental Research and Quality Assurance (Contract No. R-825404-01-0). The content of this publication has not been subject to review by either organization, and does not necessarily represent the views of either the foundation or the agency. No endorsement should be inferred.

References

1. Foy, C.L. In *Pesticide Formulation and Adjuvant Technology*; Foy, C.L.; Pritchard, D.W. Eds.; CRC Press: Boca Raton, FL, 1996; pp 323-352.
2. Hochberg, E.G. In *Pesticide Formulation and Adjuvant Technology*; Foy, C.L.; Pritchard, D.W., Eds.; CRC Press: Boca Raton, FL, 1996; pp 203-208.
3. Rosen, M.J., *Surfactants and Interfacial Phenomena*; John Wiley & Sons: New York, NY, 1989.
4. Kile, D.E.; Chiou, C.T. *Environ. Sci. Technol.* **1989**, *23*, 832-838.
5. Pennell, K.D.; Abriola, L.M.; Weber, Jr., W.J. *Environ. Sci. Technol.* **1993**, *27*, 2332-2340.
6. Karagunduz, A. *Influence of Surfactants on the Sorption and Transport of Contaminants in Saturated and Unsaturated Soils*; Ph.D. Dissertation, Georgia Institute of Technology: Atlanta, GA, 2002.
7. Pennell, K.D.; Adinolfi, A.M.; Abriola, L.M.; Diallo, M.S. *Environ. Sci. Technol.* **1997**, *31*, 1382-1389.
8. Jafvert, C.D.; Van Hoof, P.L.; Heath, J.C. *Wat. Res.* **1994**, *28*, 1009-1017.
9. Diallo, M.S.; Abriola, L.M.; Weber, W.J., Jr. *Environ. Sci. Technol.* **1994**, *28*, 1829-1837.
10. Clunie, J.S.; Ingram, B.T. In *Adsorption from Solution at the Solid/Liquid Interface*; Parkitt, G.D.; Rochester, C.H., Eds.; Academic Press: New York, NY, 1983, pp 105-152.
11. Adeel, Z.; Luthy, R.G. *Environ. Sci. Technol.* **1995**, *29*, 1032-1042.
12. Taylor, T.P.; Pennell, K.D.; Abriola, L.M.; Dane, J.H. *J. Contam. Hydrol.* **2001**, *48*, 325-350.
13. Ko, S.-O.; Schlautman, M.A.; Carraway, E.R. *Environ. Sci. Technol.* **1998**, *32*, 2769-2775.
14. Lee, J.-F.; Liao, P.-M.; Kuo, C.-C.; Yang, H.-T.; Chiou, C.T. *J. Coll. Inter. Sci.* **2000**, *229*, 445-452.
15. Edwards, D.A.; Lui, Z.; Luthy, R.G. *J. Environ. Eng. ASCE* **1994**, *120*, 23-41.
16. Sun, S.; Inskeep, W.P.; Boyd, S.A. *Environ. Sci. Technol.* **1995**, *29*, 2929-2935.
17. Pennell, K.D.; Pavlostathis, S.G.; Karagunduz, A.; Yeh, D.H. In *Chemicals in the Environment: Fate, Impacts and Remediation*, Lipnik, R.L.; Mason, R.P., Phillips, M.L., Pittman, Jr., C.U., Eds.; American Chemical Society: Washington, DC, 2002, pp. 449-466.
18. Rouse, J.D.; Sabatini, D.A.; Suflita, J.M.; Harwell, J.H. *CRC Crit. Rev. Environ. Sci. Technol.* **1994**, *24*, 325-370.
19. Demond, A.H.; Desai, F.N.; Hayes, K.F. *Water Resour. Res.* **1994**, *30*, 333-342.
20. Karagunduz, A.; Pennell, K.D.; Young, M.H. *Soil Sci Soc. Am. J.* **2001**, *65*, 1392-1399.
21. van Genuchten, M.Th. *Soil Sci Soc. Am. J.* **1980**, *44*, 892-898.

Indexes

Author Index

Agüera, A., 113
Arthur, E. L., 155
Aust, Steven D., 3
Belden, J. B., 155
Bondarenko, S., 51
Chen, Wilfred, 25
Clark, B. W., 155
Clay, David E., 199
Clay, Sharon A., 199
Coats, J. R., 155
Dungan, Robert S., 169
Ernst, Frederick F., 169
Farrell, James, 99
Feng, Yucheng, 15
Gan, J., 51, 213
Goh, K. S., 213
Guo, Mingxin, 169
Hartmann, Alain, 141
Haver, D. L., 213
Henderson, K. L., 155
Kabashima, J. N., 213
Karagunduz, Ahmet, 231
LeBlanc, Ronald, 99
Lee, S. J., 213
Lemley, A. T., 65
Malato, S., 113

Martin-Laurent, Fabrice, 141
Moorman, Thomas B., 199
Mulchandani, Ashok, 25
Papiernik, Sharon K., 169
Pennell, Kurt D., 231
Phillips, T. A., 155
Pittman, Charles U., Jr., 181
Radosevich, Mark, 129
Roberts, Charles G., 85
Saltmiras, D. A., 65
Shimazu, Mark, 25
Soulas, Guy, 141
Stahl, James D., 3
Swaner, Paul R., 3
Topp, Edward, 141
Tuovinen, Olli H., 129
Wackett, Lawrence P., 37
Wang, Jiankang, 99
Wang, Q., 65
Wu, Jiejun, 85
Wu, L. S., 213
Yates, Scott R., 169
Young, Michael H., 231
Zheng, Wei, 169
Zhu, Peter C., 85

Subject Index

A

Acetochlor
  chemical structures, 201f
  replacing alachlor, 200
Activity
  directed evolution, 32–33
  site-directed mutagenesis, 31–32
Acute toxicity test, *Vibrio fisheri*, 59
Adjuvants. *See* Surfactant adjuvants
Advanced oxidation processes (AOP), OH radical production, 114–115
Affinity tags, organophosphorus hydrolase (OPH) immobilization, 28–29
Agriculture, pesticides, 26, 114
Agrobacterium radiobacter J14a, atrazine-degrading, 148
Air quality, fumigant compounds, 170
Alachlor
  adsorption to soil and aquifer material, 204–205
  aerobic and anaerobic conditions in soil and aquifer, 208f
  bacterial EC_{50} values before and after reaction with thiosulfate salts, 60t
  chemical structures, 201f
  degradation in aerobic vs. anaerobic conditions, 210
  degradation in column, 158–159
  degradation model, 204
  dehalogenation reaction between, and thiosulfate salts, 57f
  dissipation in aqueous solutions, 55f
  effect of ammonium thiosulfate on leaching, 63f
  metabolites under aerobic and anaerobic conditions, 206, 209t, 210
  mineralization, 206
  organic carbon content and partition coefficients, 207t
  radioactive regions of thin layer chromatographic plates, 209t
  relative mobility (Rf value), 205
  relative reactivity, 57, 58f
  sorption and degradation in aquifers, 210
  structures of metabolites, 200, 202f
  transformation, 56f
Aldehydes, oxidation, 89
Aliphatic fluoro compounds, defluorination, 192–193
Ammonia, o-phthalaldehyde (OPA) precipitation, 88
Ammonium thiosulfate (ATS)
  bacterial EC_{50} values before and after reaction with thiosulfate salts, 60t
  depletion of fumigants in root zone, 173, 176–177
  detoxifying halogenated fumigants, 170
  subsurface application, 177
  See also Fumigants; Thiosulfate salts
Anodic Fenton treatment (AFT)
  AFT kinetic model, 72, 74
  analytical methods, 69–70
  anion exchange membrane AFT, 77, 79f
  apparatus of membrane AFT, 78f
  application to pesticides, 66–67
  atrazine degradation, 70, 72
  biodegradability, 80, 82
  carbamate pesticides, 80, 81f
  carbaryl degradation data, 77, 79f
  cation exchange membrane AFT, 77, 79f
  chemicals, 67

competitive kinetics, 77, 80
degradation of 2,4-D, 72, 74, 75*f*
diazinon degradation, 74, 76*f*
direct and indirect methods, 80, 81*f*
ethylene thiourea (ETU) degradation, 70
hydroxyl radical reaction rate constants, 81*f*
improvement to electrochemical Fenton treatment, 66, 70
materials and methods, 67, 69–70
membrane AFT, 77
pesticide concentration changes during AFT, 72
pesticide structures, 68*f*
proposed degradation pathway for atrazine, 73*f*
structures of ETU and degradation products, 71*f*
treatment system, 67, 69
Aquatic systems, atrazine degradation, 134
Aquifers
herbicide metabolites, 200, 203
materials and methods, 204–205
sorption and degradation of alachlor, 210
vulnerability to herbicide contamination, 203
Atrazine
abundance of degrading bacteria, 135*t*
adsorption to soil and aquifer material, 204–205
Agrobacterium radiobacter, 148
anaerobic conditions, 148–149
anodic Fenton treatment (AFT), 70, 72
aquatic and wetland systems, 134
bacterial degradation mechanisms, 142–143, 146
biologically mediated dehalogenation, 130–131
catabolic genes in environment, 136
chlorohydrolase activity, 41
Clavibacter strain, 146–147
contamination, 141–142
degradation, 200
degradation in column, 158–159
degradation potential in natural environments, 132–134
degrading microorganisms, 144*t*, 145*t*
Escherichia coli, 147
evolution of metabolic pathway, 42–45
expression of atzB in *Pseudomonas* ADP as function of medium pH, 151*f*
focus in phytoremediation, 156
hydrolytic attack, 143, 146
lower biodegradation pathway, 132*f*
magnetic capture hybridization (MCH), 149–150
metabolism to cyanuric acid, 40–42
mineralization, 136, 146, 206
monitoring bioremediating bacterial in soils, 149–150
most probable number (MPN) method, 149
nitrogen source, 42–43
Nocardioides C190, 148
organic carbon content and partition coefficients, 207*t*
oxidative attack, 142–143
PCR-denaturing gradient gel electrophoresis (DGGE) of soil community, 137*f*
percentage of applied with mulberry trees as vegetation, 160, 161*f*
phylogenetics of known, metabolizing bacteria, 135–136
proposed degradation pathway for AFT, 73*f*
Pseudaminobacter C147, 148
Pseudomonas strain, 146–148
relative mobility (Rf value), 205
soil, 132–134
soil bioremediation using atrazine-degrading bacteria, 146–149
upper biodegradation pathway, 131*f*
use, 200

weed control, 130
See also s-Triazine herbicides

B

Bacteria
  abundance of atrazine-degrading, 135*t*
  phylogenetics of known atrazine metabolizing, 135–136
  3,5,6-trichloro-2-pyridinol (TCP)-degrading, 17–18
Basins, slowing runoff flow, 220
Batch experiments, triclosan oxidation, 105–108
Bendiocarb, structure, 68*f*
Best management practices (BMPs)
  arrangement of BMPs and location of sampling points, 217*f*
  polyacrylamide (PAM), 219
  reducing runoff output and pesticide load, 216, 219–220
  sediment removal by BMPs, 222, 224–225
  sediment trap, pond and basins, 220
  vegetative strip, 220
  See also Pesticide runoff
Bifenthrin
  concentration in surface soils from nursery, 223*t*
  LC_{50} values for red and imported fire ant (RIFA) control, 216*t*
  quantification method, 221
  red and imported fire ant (RIFA) infestation, 214–215
  reduction in runoff, 227*f*
  reduction in runoff along runoff path, 226*t*
  runoff constituent, 215
Big bluestem, pendimethalin recovery from soil vegetated with, 164–165
Bioavailability
  earthworm test, 162
  lettuce seedling test, 163
  pendimethalin by earthworm uptake and lettuce seedling growth, 164*f*
  pesticide residues, 162–163
Biodegradability, carbamate pesticides, 80, 82
Biodegradation
  lower s-triazine pathway, 132*f*
  photolysis products of 3,5,6-trichloro-2-pyridinol (TCP), 20, 23
  upper s-triazine pathway, 131*f*
1,1-Bis(p-chlorophenyl)-2,2,2-trichloroethane (DDT), aqueous solubility as function of surfactant concentration, 235
Boron-doped diamond (BDD) electrodes, 103–104
  See also Triclosan

C

California Department of Pesticide Regulation (CDPR), nursery runoff, 214
Cannizzaro reactions
  advantages, 92
  disproportionation for o-phthalaldehyde (OPA) neutralization, 89, 90*f*
  final product in OPA neutralization, 93*f*
  mechanism from HPLC/MS results, 93*f*
  possible routes of OPA neutralization, 91*f*
  stepwise mechanism, 92*f*
Carbamate pesticides
  biodegradability, 80, 82
  hydroxyl radical reaction rate constants, 81*f*
Carbaryl
  anodic Fenton treatment (AFT), 77, 79*f*
  structure, 68*f*
Carbofuran, structure, 68*f*

Carbon tetrachloride, reduction, 184–185
Catabolic genes, atrazine, in environment, 136
Cellobiose dehydrogenase (CDH)
 direct and mediated oxidation and reduction reactions, 7f
 lignin degradation, 6, 8
 See also White-rot fungi
Cetyltrimethylammonium bromide (CTAB), effects on soil water pressure-saturation relationships, 241–242
Chemical vapor deposition, boron-doped diamond (BDD) electrodes, 103–104
Chemical warfare agents, destruction, 191, 192
Chloracetamide compounds
 alachlor, acetochlor, and metolachlor, 200, 201f
 degradation in soil, 200
 structures of alachlor metabolites, 202f
 See also Acetochlor; Alachlor; Metolachlor
Chlorinated aliphatic hydrocarbons (CAHs), structure, 182
5-Chloro-2-(2,4-dichlorophenoxy)-phenol. See Triclosan
2-Chloro-4-ethylamino-6-isopropylamino-s-triazine. See Atrazine
2-Chloro-4-hydroxy-6-amino-2-atrazine, dechlorination, 42
Chloroacetanilide, disappearance in thiosulfate solutions, 58f
Chlorofluorocarbons (CFCs), remediation, 190
Chloropicrin
 bacterial EC_{50} values before and after reaction with thiosulfate salts, 60t
 half-lives with and without ammonium thiosulfate, 59t
 partial replacement for methyl bromide, 170
 reaction rate constant and regression coefficient, 55t
Chlorpyrifos
 degradation, 16
 LC_{50} values for red and imported fire ant (RIFA) control, 216t
 See also 3,5,6-Trichloro-2-pyridinol (TCP)
CIDEX® OPA solution. See o-Phthalaldehyde (OPA)
Clavibacter strain, atrazine-degrading, 146–147
Competitive kinetics
 anodic Fenton treatment (AFT), 77, 80
 hydroxyl radical reaction rate constants, 81f
Compound parabolic collectors (CPCs), solar photocatalysis, 117
Contamination. See Phytoremediation; Remediation
Cyanuric acid
 atrazine metabolism to, 40–42
 enzymatic hydrolysis, 40f

D

2,4-D
 anodic Fenton treatment (AFT), 72, 74
 degradation with different delivery rates of Fenton reagent, 75f
 structure, 68f
Dakin reaction
 aromatic aldehydes, 94
 electron-donating groups, 95f
 example, 94f
 oxidation of aromatic aldehydes, 93
 requirements, 95
Dechlorination
 activity of different metals and halides during solvated electron, 188t
 aromatic hydrocarbons, 184
 chlorinated organics, 182–183

2-chloro-4-hydroxy-6-amino-s-
triazine (CAOT), 42
dissociative electron transfer, 184,
185
efficiencies of Na and Ca, 186–187
excess water, 185–187
halogenated organic compounds,
52
hydrogen generation, 186
minimum Na for complete, 186,
187t
reduction of carbon tetrachloride,
184–185
transfer of solvated electron to
chlorinated aliphatic hydrocarbon,
184
See also Remediation
Defluorination, aliphatic fluoro
compounds, 192–193
Degradation
lignin, 4–5
microbial, of organophosphates, 27–
28
pentachlorophenol (PCP), 8–10
trichloroethylene, 11
trinitrotoluene (TNT), 10, 12f
See also Anodic Fenton treatment
(AFT); 3,5,6-Trichloro-2-pyridinol
(TCP); White-rot fungi
Dehalogenation
activity of metals and halides during
solvated electron, 188t
aerobic, of hydrocarbons by white-
rot fungi, 12f
reaction between alachlor and
thiosulfate salts, 57f
trichloroethylene, 11
Deoxyribonucleic acid (DNA)
shuffling, directed evolution, 32–33
Detoxification
halogenated organic compounds
(HOC), 59–60
pentachlorophenol (PCP), 8–10
solar, of pesticides, 118–123
trinitrotoluene (TNT), 10, 12f

See also Halogenated organic
compounds (HOC);
Organophosphates;
Phytoremediation; White-rot fungi
Dextran, *o*-phthalaldehyde (OPA)
neutralization, 88
Diazinon
anodic Fenton treatment (AFT), 74
LC_{50} values for red and imported fire
ant (RIFA) control, 216t
structure, 68f
variation of oxidation rate parameter
and electrolysis current efficiency
with NaCl concentration, 76f
1,2-Dichlorobenzene, structure, 182
1,3-Dichloropropene (1,3-D)
bacterial EC_{50} values before and after
reaction with thiosulfate salts, 60t
concentration in soil gas, 174f
fumigant dissipation rate, 177t
half-lives with and without
ammonium thiosulfate, 59t
mass remaining in replicate
mesocosms after ammonium
thiosulfate/water, 176f
partial replacement for methyl
bromide, 170
reaction rate constant and regression
coefficient, 55t
Dioxacarb, structure, 68f
Dioxins
Na/NH$_3$ treatment, 188t
structure, 182
Directed evolution, DNA shuffling,
32–33
Disappearance, pesticides, 119–121
Disposal methods, organophosphate
compounds, 26
Disposal problems, triclosan, 100–101
Dissipation model, first-order for
halogenated fumigants, 176
Diuron, disappearance and
mineralization, 120f
DNA shuffling, directed evolution,
32–33

E

Earthworm assay
  bioavailability, 162
  bioavailability of pendimethalin, 164f
Electrochemical oxidation
  hydroxyl radicals, 102–103
  organic compounds, 101–103
  See also Triclosan
Electrodes, boron-doped diamond (BDD), 103–104
Emissions-reduction strategies, fumigants, 171
Environmental pollution, global problem, 4
Environmental Protection Agency (EPA), Toxics Release Inventory, 195
Escherichia coli
  atrazine-degrading, 147
  expressing organophosphorus hydrolase (OPH) and cellulose-binding domain (CBD), 31
Ethylene thiourea (ETU)
  anodic Fenton treatment (AFT), 70
  degradation products, 71f
  structure, 71f
Explosives, remediation of soil, 192t

F

Fenobucarb, structure, 68f
Fenoxycarb, LC_{50} values for red and imported fire ant (RIFA) control, 216t
Fenton processes, production of OH radicals, 115
Fenton treatment method
  reaction, 66
  See also Anodic Fenton treatment (AFT)
Ferricyanide, detoxification, 10, 12f
Flow-through reactor, triclosan oxidation, 108–111
Fumigant emissions reduction, methyl bromide, 60–61
Fumigants
  air quality, 170
  concentration of propargyl bromide and cis-1,3-dichloropropene (1,3-D) in soil gas, 174f
  concentrations in root zone, 173
  concrete mesocosms, 171
  decrease in propargyl bromide and methyl isothiocyanate (MITC) in root zone with time, 178f
  degradation at soil surface, 170
  depletion in root zone by ammonium thiosulfate (ATS) application, 173, 176–177
  detoxifying halogenated, 170
  dimensions of mesocosms, 172f
  emissions-reduction strategies, 171
  first-order dissipation model, 176
  initial soil gas concentrations, 173
  location of soil gas samplers in xy plane, 172f
  mass remaining in replicate mesocosms after ATS/water application, 176f
  methods, 171, 173
  methyl bromide (MeBr), 170
  normalized soil gas concentrations for propargyl bromide in mesocosm, 175f
  properties, 170
  rate of fumigant dissipation, 177t
  soil gas concentration measurements, 171, 173
  subsurface application, 177

G

Genes, atrazine catabolic, in environment, 136
Glycine, *o*-phthalaldehyde (OPA) neutralization, 86–87
Green neutralization, *o*-phthalaldehyde (OPA), 88

H

Halogenated organic compounds (HOC)
  advantages of thiosulfate salts, 53
  alachlor transformation, 56f
  application examples, 60–62
  bacterial EC_{50} values of fumigants and herbicides, 60t
  dehalogenation reaction between alachlor and thiosulfate, 57f
  detoxification, 59–60
  disappearance of chloroacetanilide herbicides in thiosulfate, 58f
  dissipation of alachlor in aqueous solutions, 55f
  dissipation of fumigants in thiosulfate solutions, 54f
  effect of ammonium thiosulfate on alachlor and propachlor leaching, 63f
  enhanced transformation in soil, 58–59
  fumigant emissions reduction, 60–61
  half-lives of fumigants with and without ammonium thiosulfate (ATS), 59t
  propargyl bromide transformation, 56f
  reaction pathways, 55, 57
  reaction rate constants and regression coefficient of fumigants and ATS, 55t
  reaction routes, 52–53
  relative reactivity, 57, 58f
  remediation in ground water and polluted soil, 52
  soil remediation, 61–62
  thiosulfate salts to dehalogenate HOCs, 53
  transformation kinetics in solution, 53–54
  uses, 52
Herbicides
  adjuvants, 232
  extraction method for degradates, 205
  materials and methods, 204–205
  metabolites in aquifers, 200, 203
  relative reactivity, 57, 58f
  toxicological properties, 203
  See also s-Triazine herbicides
Hexachlorobenzene (HCB)
  aqueous solubility as function of surfactant concentration, 235
  effect of surfactant on HCB sorption, 240f
  Na/NH_3 treatment, 189
  phase distribution of HCB in surfactant-Appling soil system, 241f
History, s-triazine herbicide, 39–40
Hospital instrument processing, *o*-phthalaldehyde (OPA) solution, 86
Hydramethylnon, LC_{50} values for red and imported fire ant (RIFA) control, 216t
Hydrogen peroxide
  Dakin reaction, 93–95
  ideal neutralization of *o*-phthalaldehyde (OPA), 96f
  neutralization by oxidation, 93–96
  OPA neutralization, 94f
Hydrolytic attack, bacterial degradation of atrazine, 143, 146
Hydrophile–lipophile balance (HLB), definition, 233
Hydrophobic organic compound (HOC). *See* Surfactant adjuvants
Hydroxyatrazine
  generation, 41
  metabolism by hydroxyatrazine amidohydrolase, 42
Hydroxyl radicals
  Fenton reagent, 115
  production in advanced oxidation processes (AOP), 114–115
Hydrogen, generation, 186

I

Ice-nucleation protein (INP), surface expression, 30–31
Imidacloprid, disappearance and mineralization, 120f, 121
Immobilization, organophosphorus hydrolase (OPH), by affinity tags, 28–29
Industrial wastewater
 designing treatment plan for, 122–123
 effect of sodium chloride on TCP removal, 19f
 solar detoxification, 118
 treatment of 3,5,6-trichloro-2-pyridinol (TCP)-containing, 18, 20
Iodomethane, soil fumigant, 170

K

Kinetic model, anodic Fenton treatment (AFT), 72, 74
Kinetics, anodic Fenton treatment (AFT), 77, 80

L

Laccase
 direct and mediated oxidation and reduction reactions, 7f
 one electron oxidation, 6
 See also White-rot fungi
Langmuir-Hinselwood, kinetic model, 119
Leachate, pesticide recovery in, during phytoremediation, 161, 162f
Lettuce seedling
 bioavailability, 163
 bioavailability of pendimethalin, 164f
Lignin
 degradation by white-rot fungi, 4–5
 See also White-rot fungi
Lindane, structure, 181

M

Magnetic capture hybridization (MCH), soil mineralizing atrazine, 149–150
Malathion, LC50 values for red and imported fire ant (RIFA) control, 216t
Mechanism, Cannizzaro reaction, 92f, 93f
Membrane anodic Fenton treatment (AFT)
 advance in AFT, 77
 apparatus, 78f
 changes in degradation rate for carbaryl and electrolysis voltage with NaCl concentration, 79f
Mesocosms
 dimensions, 172f
 fumigants, 171
Metabolism
 atrazine to cyanuric acid, 40–42
 s-triazine herbicide, 39–40
 3,5,6-trichloro-2-pyridinol (TCP), 18, 19f
Methylation, pentachlorophenol (PCP), 8, 9f
Methyl bromide
 bacterial EC_{50} values before and after reaction with thiosulfate salts, 60t
 fumigant emission reduction, 60–61
 half-lives with and without ammonium thiosulfate, 59t
 reaction rate constant and regression coefficient, 55t
Methyl iodide
 bacterial EC_{50} values before and after reaction with thiosulfate salts, 60t
 dissipation in thiosulfate, 54f
 half-lives with and without ammonium thiosulfate, 59t

reaction rate constant and regression coefficient, 55t
Methyl isothiocyanate (MITC)
  decrease in root zone with time, 178f
  fumigant dissipation rate, 177t
  mass remaining in replicate mesocosms after ammonium thiosulfate/water, 176f
  partial replacement for methyl bromide, 170
Metolachlor
  chemical structures, 201f
  concentration after 250 days remediation, 159f
  focus in phytoremediation, 156
  movement within soil column during phytoremediation, 161, 162f
  percentage of applied with mulberry trees as vegetation, 160, 161f
  replacing alachlor, 200
Micellar solubilization
  aqueous solubility of 1,1-bis(p-chlorophenyl)-2,2,2-trichloroethane (DDT) and hexachlorobenzene (HCB), 235
  change in surfactant monomer and micelle mass fraction vs. total surfactant concentration, 234f
  critical micelle concentration (CMC), 234
  effect of surfactants on hydrophobic organic compound (HOC) solubility, 236
  surfactant micelles, 234–235
  weight solubilization ratio (WSR), 235–236
  See also Surfactant adjuvants
Microbial degradation
  organophosphates, 27–28
  s-triazines, 130–136
  See also Atrazine
Microorganisms, atrazine-degrading, 144t, 145t
Microtox® assay
  reaction products of triclosan, 109, 111
  toxicity of triclosan solutions, 104–105
Mineralization
  atrazine, 136, 146, 206
  pesticides, 119–121
Mirex, structure, 181
Mitigation. *See* Pesticide runoff
Model
  first-order dissipation, for halogenated fumigants, 176
  kinetic, of anodic Fenton treatment (AFT), 72, 74
Most probable number (MPN) methodology
  atrazine-degraders, 132–133
  bioremediating bacteria in soils, 149–150
Mulberry trees
  phytoremediation potential, 160, 161f
  See also Phytoremediation

N

Neurotoxins. *See* Organophosphates
Neutralization. *See* o-Phthalaldehyde (OPA)
Nitrogen heterocyclic compounds, exposure of living cells, 38
Nocardioides C190, atrazine-degrading, 148
Non-concentrating solar collectors, solar photocatalysis, 117
Non-halogenated molecules, reductive remediation, 190–191
Nursery production
  bifenthrin against red and imported fire ant (RIFA), 214–215
  industry, 214
  mid-size commercial nursery as field site, 214
  pesticide fate and safety, 215
  pyrethroid insecticides, bifenthrin and permethrin, 215
  See also Pesticide runoff

O

Organophosphates
  crystal structure of organophosphorus hydrolase (OPH), 32
  directed evolution, 32–33
  disposal methods, 26
  DNA shuffling, 32–33
  enzyme detoxification, 28–33
  genetically engineered *Escherichia coli* cell, 31
  ice-nucleation protein (INP), 30–31
  kinetic properties of organophosphorus hydrolase (OPH), 27*t*
  microbial degradation, 27–28
  modifications of specificity and activity, 31–33
  neurotoxins, 26
  OPH description, 27
  OPH immobilization by affinity tags, 28–29
  site-directed mutagenesis, 31–32
  surface expression of OPH, 30–31
  toxicity and usage, 26
  truncated ice-nucleation protein (INPNC), 31
  whole cell detoxification of OP neurotoxins, 29–30
Organophosphorus hydrolase (OPH)
  crystal structures, 32
  description, 27
  immobilization by affinity tags, 28–29
  kinetic properties, 27*t*
  site-directed mutagenesis, 31–32
  surface expression of OPH, 30–31
Oxidation
  aldehydes, 89
  pentachlorophenol (PCP), 9*f*
  See also Hydrogen peroxide; *o*-Phthalaldehyde (OPA)
Oxidative attack, bacterial degradation of atrazine, 142–143

P

Pendimethalin
  bioavailability, 163, 164*f*
  concentration after 250 days remediation, 159*f*
  focus in phytoremediation, 156
  percentage remaining in soil with and without vegetation, 158*f*
  recovery in soil vegetated with switchgrass, big bluestem, and prairie grass mixture, 165*f*
  See also Phytoremediation
Pentachlorophenol (PCP)
  cycle of oxidative, reductive, and methylation reactions, 9*f*
  detoxification and degradation, 8–10
  methylation of PCP, 9*f*
  structure, 182
  See also White-rot fungi
Permethrin
  concentration in surface soils from nursery, 223*t*
  quantification method, 221
  reduction in runoff, 227*f*
  reduction in runoff along runoff path, 228*t*
  runoff constituent, 215
Peroxidases
  direct and mediated oxidation and reduction reactions, 7*f*
  lignin degradation, 5–6
  See also White-rot fungi
Peroxide. *See* Hydrogen peroxide
Persistent organic compounds, classification, 114
Pesticide residues, bioavailability, 162–163
Pesticide runoff
  arrangement of best management practices (BMPs) and location of sampling points, 217*f*
  best management practices (BMPs), 216, 219–220

bifenthrin against red and imported fire ant (RIFA) infestation, 214–215
concentrations of bifenthrin and permethrin in surface soils, 223t
experimental, 215–221
LC$_{50}$ values of insecticides for control of RIFA, 216t
locations for surface soil sampling at nursery site, 218f
monthly runoff output, 219f
nursery production for pesticide fate and safety, 215
percentages of reduction in sediment level in runoff water, 225t
pesticide removal by BMPs, 225–226, 228–229
polyacrylamide (PAM) for improving, 219
reduction in bifenthrin in runoff at different locations, 227f
reduction in bifenthrin level in runoff along runoff path, 226t
reduction in permethrin in runoff along runoff path, 228t
reduction in permethrin in runoff at different locations, 227f
reduction in suspended solid content of runoff, 224f
sampling and analysis, 220–221
sediment removal by BMPs, 222, 224–225
sediment trap, pond and basins slowing down runoff flow, 220
site characterization and survey for pesticide sources, 215–216
sources of pesticides in nursery runoff, 221–222
vegetative strip, 220
Pesticides
agriculture, 114
dechlorination, 182–183
destruction methods, 182
disappearance and mineralization, 120f
fate of radio-labeled, within prairie grass-soil system, 163–164
future for photocatalysis, 123–124
kinetic model, 119
modern agriculture, 26
movement within soil column during phytoremediation, 161, 162f
nitrogen heterocyclic, 38t
persistent organic compound classification, 114
perspective, 195–196
remediation, 189
solar detoxification, 118–123
structures, 181–182
water contaminants, 66
See also Anodic Fenton treatment (AFT); Halogenated organic compounds (HOC); Organophosphates
Phanerochaete chrysosporium. *See* White-rot fungi
Photocatalysis
compound parabolic collectors (CPCs), 117
detoxication by solar, 115
future outlook for pesticides, 123–124
laboratory research, 116–118
non-concentrating solar collectors, 117
pilot plant scheme, 116f
schematic of CPCs, 117f
Photodegradation
proposed pathway for 3,5,6-trichloro-2-pyridinol (TCP), 22f
TCP, 20
TCP experiments, 17
TCP upon exposure to ultraviolet (UV) light, 21f
o-Phthalaldehyde (OPA)
advantages of Cannizzaro reactions, 92
advantages of CIDEX® OPA solution, 86
ammonia precipitating OPA, 88

Cannizzaro final product by GC/MS, 93f
Cannizzaro mechanism by HPLC/MS results, 93f
Cannizzaro reaction for neutralization by sodium hydroxide, 89, 92
Dakin reaction, 93-95
hospital instrument processing, 86
ideal green method, 88
ideal neutralization by hydrogen peroxide, 96f
neutralization by glycine, 86-87
neutralization by scavenging on dextran, 88
neutralization by scavenging on silica support, 87
neutralization by sodium bisulfite, 86
neutralization with hydrogen peroxide by oxidation, 93-96
possible neutralization products by redox or disproportionation reactions, 90f
possible routes for Cannizzaro reactions, 91f
reduction by sodium borohydride, 87
starch support, 88
stepwise mechanism of Cannizzaro reactions, 92f
Phylogenetics, atrazine metabolizing bacteria, 135-136
Phytoremediation
bioavailability of pendimethalin, 163
bioavailability of pesticide residues, 162-163
concentration of metolachlor and pendimethalin after 250 days, 159f
considerations, 165-166
definition, 156
diversity, 156
earthworm bioavailability assay, 162
evaluating role of species type and mixture, 164-165
evaluating success using alternative endpoints, 160-163
fate of radio-labeled pesticides within prairie grass-soil systems, 163-164
lettuce seedling bioavailability assay, 163
mulberry trees, 160
percentage of applied atrazine, metolachlor, and pendimethalin recovered from columns, 161f
percentage of metolachlor recovered in leachate, 162f
percentage of pendimethalin and trifluralin remaining for vegetated and unvegetated, 158f
pesticide movement within soil column during, 161
pesticides in focus, 156
plants of interest, 156-157
potential for point-source pesticide contamination, 156
prairie grasses in column study, 158-159
prairie grasses in microplot study, 157-158
recovered pendimethalin in big bluestem, switchgrass, and mixed prairie grasses, 165f
techniques, understanding and improving, 163-165
Plasma membrane redox potential, metabolism of chemicals, 10, 12f
Pollution, global problem, 4
Polyacrylamide (PAM), reducing runoff, 219
Polychlorinated biphenyls (PCBs)
destruction in oils using Na/NH$_3$, 189t
destruction in soils using Na/NH$_3$ or Ca/NH$_3$, 183t
incineration, 182
Na/NH$_3$ treatment, 188t
structure, 181, 182
Polynuclear aromatic hydrocarbon (PNAs), reductive remediation, 190-191
Pond, slowing runoff flow, 220

Prairie grasses
  column study of phytoremediation, 158–159
  fate of radio-labeled pesticides within prairie grass-soil system, 163–164
  microplot phytoremediation study, 157–158
  pendimethalin recovery from soil vegetated with, 164–165
  See also Phytoremediation
Promecarb, structure, 68*f*
Propachlor
  effect of ammonium thiosulfate on leaching, 63*f*
  relative reactivity, 57, 58*f*
Propargyl bromide
  bacterial EC_{50} values before and after reaction with thiosulfate salts, 60*t*
  concentration in soil gas, 174*f*
  decrease in root zone with time, 178*f*
  dissipation in thiosulfate, 54*f*
  fumigant dissipation rate, 177*t*
  half-lives with and without ammonium thiosulfate (ATS), 59*t*
  mass remaining in replicate mesocosms after ATS/water, 176*f*
  normalized soil gas concentrations, 175*f*
  reaction rate constant and regression coefficient, 55*t*
  soil fumigant, 170
  transformation, 56*f*
Pseudaminobacter C147, atrazine-degrading, 148
Pseudomonas species
  atrazine-degrading, 146–148
  expression of atzB as function of medium pH, 151*f*
  metabolism of 3,5,6-trichloro-2-pyridinol (TCP) by resting cells, 19*f*
  TCP-degrading, 17–18
Pyriproxyfen, LC_{50} values for red and imported fire ant (RIFA) control, 216*t*

Q

Quinones, detoxification, 10, 12*f*

R

Radio-labeled pesticides, fate within prairie grass-soil system, 163–164
Reaction pathways, halogenated organic compounds with thiosulfate salts, 55, 57
Reactivity, herbicides, 57, 58*f*
Red and imported fire ant (RIFA) infestation, bifenthrin, 214–215
Reduction
  destruction of polychlorinated biphenyls (PCBs), 183*t*
  nitro- and nitrate containing organic compounds, 191
  pentachlorophenol (PCP), 9*f*
  polychlorinated biphenyls (PCBs), 182
  sodium borohydride, 87
  solvated electrons, 183
Reductive dechlorination, halogenated organic compounds, 52–53
Remediation
  CH_3CCl_3 soils, 194*t*
  chlorofluorocarbons (CFCs), 190
  defluorination of aliphatic fluoro compounds, 192–193
  destruction of chemical warfare agents, 191, 192*t*
  destruction of polychlorinated biphenyls (PCBs), 189*t*
  Na consumption vs. concentration of chlorinated contaminant in soil, 193–195
  Na/NH_3 treatment, 188*t*
  pesticide-contaminated soils using Na/NH_3 in mobile unit, 190*t*
  pesticides, 189
  polynuclear aromatic hydrocarbons (PNAs), 190–191

reductive, of nonhalogenated molecules, 190–191
sodium consumption vs. concentration, 193–195
soil, with thiosulfate salts, 61–62
tetrachloroethylene-contaminated soils, 194t
See also Phytoremediation
Root zone
decrease of propargyl bromide and methyl isothiocyanate (MITC) with time, 178f
fumigant depletion, 173, 176–177
See also Fumigants
Runoff. *See* Pesticide runoff

S

Sediment trap, slowing runoff flow, 220
Silica support, *o*-phthalaldehyde (OPA), 87
Site-directed mutagenesis, organophosphorus hydrolase (OPH), 31–32
Sodium bisulfite, *o*-phthalaldehyde (OPA) neutralization, 86
Sodium borohydride, reduction reaction, 87
Sodium chloride
effect of concentration on 3,5,6-trichloro-2-pyridinol (TCP) removal, 19f
industrial wastewater treatment, 18, 20
Sodium consumption, factors affecting, 193–195
Sodium hydroxide, *o*-phthalaldehyde (OPA) neutralization, 89, 91f, 92
Sodium ions. *See* Solvated electrons
Soil
abundance of atrazine degrading bacteria, 135t
atrazine degradation, 132–134
destruction of polychlorinated biphenyls (PCBs), 183t
enhanced transformation, 58–59
hydrolytic attack for bacterial degradation of atrazine, 143, 146
monitoring abundance and activity of bioremediating bacteria, 149–150
oxidative attack of atrazine, 142–143
PCR–denaturing gradient gel electrophoresis (DGGE), 137f
remediation of contaminated with explosives, 192t
remediation of pesticide-contaminated, 190t
surfactant sorption, 237–238
See also Atrazine; Surfactant adjuvants
Soil column, pesticide movement within, during phytoremediation, 161, 162f
Soil gas
concentrations at root zone, 173
initial concentrations, 173
measuring concentrations, 173
See also Fumigants
Soil remediation
thiosulfate salts, 61–62
See also Phytoremediation
Soil water retention relationships, van Genuchten (VG) equation, 242
Solar collectors, photocatalysis, 117
Solar detoxification
commercial pesticides, 121–123
disappearance and mineralization of commercial pesticides, 122f
disappearance of pesticides, 120f
equation describing kinetics, 119
future outlook, 123–124
imidacloprid and formulation, 121
mineralization of pesticides, 120f
photocatalytic disappearance with TiO_2, 119, 121
pure pesticides, 119, 121
treatment of industrial wastewater, 118–119

treatment plant for wastewaters, 122–123
Solar photocatalysis, detoxication, 115
Solvated electrons
  activity of metals and halides during solvated electron dechlorinations, 188t
  Na or Ca for reductions, 183
Specificity
  directed evolution, 32–33
  site-directed mutagenesis, 31–32
Starch supports, o-phthalaldehyde (OPA) neutralization, 88
Surface and aquifer sediments
  alachlor mineralization, 206
  alachlor under aerobic and anaerobic conditions, 206, 208f
  atrazine mineralization, 206
  materials and methods, 204–205
  metabolites of alachlor under aerobic and anaerobic conditions, 206, 210
  organic carbon content and partition coefficients of atrazine and alachlor, 207t
  physical and chemical characteristics, 205–206
  relative mobility (Rf value), 205, 209t
  sorption and degradation of alachlor, 210
Surfactant adjuvants
  apparent soil-water distribution coefficient (K_D^*), 239–240
  aqueous solubility of 1,1-bis(p-chlorophenyl)-2,2,2-trichloroethane (DDT) and hexachlorobenzene (HCB) vs. Triton X-100 concentration, 235f
  cetyltrimethylammonium bromide (CTAB) and soil water pressure-saturation relationships, 241–242
  change in surfactant monomer and micelle mass fraction vs. total surfactant concentration, 234f
  coupled hydrophobic organic compound (HOC) and surfactant sorption, 238–240
  critical micelle concentration (CMC), 234
  effect of Tween 80 on HCB sorption by Appling soil, 240f
  effects of surfactant-induced changes on surface tension and contact angle, 242
  effects of surfactants on HOC solubility, 236
  effects of surfactants on soil water retention and flow, 241–243
  herbicide formulations, 232
  HOC phase distribution in surfactant-soil systems, 236–240
  hydrophile-lipophile balance (HLB), 233
  impact on water flow and coupled transport, 242–243
  influence of Triton X-100 pulse injection on soil water pressure in unsaturated column, 244f
  Langmuir sorption parameters for surfactant-soil systems, 238t
  measured and fitted soil water retention curves for F-70 Ottawa sand, 243f
  micellar solubilization, 233–236
  phase distribution of HCB in Tween 80–Appling soil system, 241f
  properties of representative nonionic surfactants as pesticide adjuvants, 233t
  schematic of three-phase soil system, 237f
  soil water retention (pressure-saturation) relationships, 242
  surfactant properties, 232–233
  surfactant sorption, 237–238
  van Genuchten (VG) equation, 242
  weight solubilization rate (WSR), 235–236
Switchgrass, pendimethalin recovery from soil vegetated with, 164–165

T

Tags, affinity, organophosphorus hydrolase (OPH) immobilization, 28–29
Tetrachloroethylene, structure, 182
Thiosulfate salts
 advantages, 53
 bacterial EC_{50} values of herbicides before and after reaction with, 60t
 dehalogation of halogenated organic compounds (HOC), 53
 dehalogenation reaction between alachlor and, 57f
 dissipation of fumigants in, 54f
 enhanced transformation in soil, 58–59
 fumigant emissions reduction, 60–61
 soil remediation, 61–62
 transformation of HOCs, 53–54
 See also Halogenated organic compounds (HOC)
Toxicity
 Microtox® assay for triclosan, 104–105
 triclosan, 109, 111
Toxics Release Inventory, pesticides, 195
s-Triazine herbicides
 atrazine, 42–43
 atrazine chlorohydrolase gene, 44
 atrazine metabolism to cyanuric acid, 40–42
 atzA catalyzing dehalogenation, 45
 atzD, atzE, and atzF genes, 43
 2-chloro-4-hydroxy-6-amino-s-triazine (CAOT), 42
 commercially relevant nitrogen heterocyclic pesticides, 38t
 enzymatic hydrolysis of cyanuric acid, 40f
 evolution of metabolic pathway for atrazine, 42–45
 fate, 39
 history, 39–40
 hydroxyatrazine metabolism, 42
 N-isopropylammelide isopropylaminohydrolase (AtzC), 42
 laboratory experiments emulating evolutionary pathway between triA and atzA, 44–45
 metabolism, 39–40
 metabolism of atrazine in *Pseudomonas* sp. ADP, 41f
 microbial degradation, 130–136
 origins of atzA, atzB, and atzC genes, 43
 success, 38–39
 weed control, 130
 See also Atrazine
3,5,6-Trichloro-2-pyridinol (TCP)
 biodegradation of TCP photolysis products, 20, 23
 degradation metabolite, 16
 effect of sodium chloride concentration on TCP removal, 19f
 enrichment culture techniques, 16–17
 experimental, 16–17
 isolation and characterization of TCP-degrading bacteria, 17–18
 metabolism by resting cells of *Pseudomonas* sp., 19f
 photodegradation, 20
 photodegradation experiments, 17
 photodegradation upon exposure to UV light, 21f
 proposed photodegradation pathway, 22f
 treatment of TCP-containing industrial wastewater, 18, 20
Trichloroethylene
 dehalogenation and degradation, 11
 See also White-rot fungi
Triclopyr
 degradation, 16
 See also 3,5,6-Trichloro-2-pyridinol (TCP)
Triclosan
 batch experiments, 104
 batch reactor, 105–108

boron-doped diamond (BDD) electrodes, 103–104
concentrations as function of electrolysis time, 106, 107f
cyclic voltammetry scans with BDD electrode, 105f
disposal problems, 100–101
effluent concentrations from flow-through reactor, 109f
electrochemical oxidation, 101–103
experiments in DiaCell® silicon substrate (CSEM) flow-through reactor, 104
flow-through reactor, 108–111
galvanostatic electrolysis at anodic current densities of 5 and 15 mA/cm^2, 107f
major byproducts of electrolysis, 110f
materials and methods, 104–105
Microtox® assay, 104–105
Microtox® EC_{50} values of reaction products, 109, 111
normalized toxicity and concentrations during electrolysis, 111f
oxidation by hydroxyl radicals, 102–103
removal rates at current densities of 5 and 15 mA/cm^2, 106, 108
structure, 100f
uses, 100
Trifluralin
focus in phytoremediation, 156
percentage of applied with mulberry trees as vegetation, 160, 161f
percentage remaining in soil with and without vegetation, 158f
See also Phytoremediation
Trinitrotoluene (TNT)
detoxification and degradation, 10, 12f
See also White-rot fungi
Truncated ice nucleation protein (INPNC), surface expression, 31

V

van Genuchten (VG) equation, soil water retention relationships, 242
Vegetation
role of species type and mixture in phytoremediation, 164–165
See also Phytoremediation
Vegetative strip, slowing runoff flow, 220
Vibrio fisheri, acute toxicity test, 59

W

Wastewater
designing treatment plan for, 122–123
effect of sodium chloride on 3,5,6-trichloro-2-pyridinol (TCP) removal, 19f
treatment of TCP-containing, 18, 20
Water, oxidative attack of atrazine, 142–143
Weight solubilization ratio (WSR), definition, 235–236
Wetland systems, atrazine degradation, 134
White-rot fungi
aerobic dehalogenation of hydrocarbons, 12f
cellobiose dehydrogenase, 6, 8
chemicals by fungal plasma membrane redox system, 12f
direct and mediated oxidation and reduction reactions, 7f
laccase, 6
lignin, 4–5
methylation of pentachlorophenol (PCP), 9f
oxidative, reduction, and methylation reactions for PCP degradation, 9f
pentachlorophenol detoxification and degradation, 8–10
peroxidases, 5–6
reaction possibilities, 5–8